# Managing Reality

## Book Two
## Procuring an Engineering and Construction Contract

Second edition

# Managing Reality

## Book Two
## Procuring an Engineering and Construction Contract

Second edition

**Bronwyn Mitchell and Barry Trebes**

Published by ICE Publishing, 40 Marsh Wall, London E14 9TP.

Full details of ICE Publishing sales representatives and distributors can be found at:
www.icevirtuallibrary.com/printbooksales

First edition published 2005

**Also available in this series from ICE Publishing Books**
*Managing Reality, Book One: Introduction to the Engineering and Construction Contract, Second edition.* ISBN 978-0-7277-5718-0
*Managing Reality, Book Three: Managing the Contract, Second edition.* ISBN 978-0-7277-5722-7
*Managing Reality, Book Four: Managing Change, Second edition.* ISBN 978-0-7277-5724-1
*Managing Reality, Book Five: Managing Procedures, Second edition.* ISBN 978-0-7277-5726-5

A catalogue record for this book is available from the British Library

9 8 7 6 5 4 3 2 1

ISBN 978-0-7277-5720-3

© Thomas Telford Limited 2012

ICE Publishing is a division of Thomas Telford Ltd, a wholly-owned subsidiary of the Institution of Civil Engineers (ICE).

FSC
www.fsc.org
MIX
Paper from
responsible sources
FSC® C013604

Typeset by Academic + Technical, Bristol
Printed and bound by CPI Group (UK) Ltd, Croydon CR0 4YY

# Contents

# Preface

In the preface to the first edition of *Managing Reality*, in 2005, we set out our aims and aspirations for 'Managing Reality'. These were as follows.

- To add and contribute to the body of knowledge on the use of the NEC ECC.
- To provide a book which focuses on the 'how to': how to manage and administer the ECC contract.
- To present as a five-part book-set that covers both the needs of the student professional or prospective client, through to the novice practitioner and experienced user.
- To provide a rounded view of the ECC, whatever your discipline, on both sides of the contractual relationship
- To enable everyone to realise the business benefits from using the NEC suite of contracts generally and the ECC in particular.
- *Managing Reality* does not attempt to give a legal treatise or a blow-by-blow review of each and every clause. It is intended to be complementary to other publications, which give excellent theoretical and legal perspectives.

This book is about dealing with the reality of real life projects: Managing Reality.

The feedback and support we have received since its first publication in 2005 has been universally positive and we would like to thank all of you who have bought and used *Managing Reality* since its first publication.

We have greatly enjoyed updating and working on this second edition and we hope that it continues to provide a useful body of knowledge on the use of the NEC3 ECC.

**Bronwyn Mitchell and Barry Trebes**

# Foreword

A key objective of the first edition of *Managing Reality* was to provide a five-part book to meet the needs of students, prospective clients, novice practitioners and experienced users. Satisfying such diverse needs is an ambitious objective for any text.

Does *Managing Reality* achieve its stated aim? I believe that the answer to this is a resounding 'yes'. In my view, the calibre of authorship is exceptional. All levels of and types of readership from the uninitiated to the experienced professional will derive considerable benefit from this text. Although written in a very accessible style, there is no skimping on detail or on addressing difficult issues. The worked examples are particularly helpful. *Managing Reality* should be your prime aid from the moment you are considering whether or not to use an NEC contract right through to using and operating the contract.

But *Managing Reality* is much more than simply a 'how to' guide. It seeks to deliver a clear message that NEC contracts cannot be used to their full potential unless one is prepared to ditch one's knowledge and experience of traditional contracting. For example, emphasis is placed on the fact that certainty and predictability are the hallmarks of NEC contracts. Open-ended and subjective phrases and concepts have no place in NEC contracting.

I am privileged to be associated with this second edition of *Managing Reality*. It will continue to help those who need help overcoming any reservations about using NEC contracts and re-inforce existing users in their continued use of these ground-breaking contracts.

**Professor Rudi Klein**
President, NEC Users' Group

# Acknowledgements

We would like to thank the following individuals and companies who have supported the book.

For their active participation in this book we would like to thank

- Professor Rudi Klein (SEC Group Chief Executive) for writing the Foreword
- Dr Robert N. Hunter of Hunter and Edgar Edinburgh for his thoughts and suggested revisions for this second edition
- Gavin Jamieson, the Senior Commissioning Editor, for his enthusiasm and patience
- colleagues at Mott MacDonald
- everyone who has given feedback on the book since 2005.

And our continued gratitude to those who provided support and input into the first edition of *Managing Reality*.

- Mike Attridge, of Ellenbrook Consulting, who reviewed the book on behalf of the authors.
- David H. Williams who provided guidance and support in the development of the book.
- Everyone at Needlemans Limited Construction Consultants (now part of the Mott MacDonald Group).
- Everyone at MPS Limited with whom Needlemans Limited worked to develop the first web based management system for the NEC in 2000.

Finally, we would like to thank our family and friends for their on-going support, understanding and patience.

# Series contents

The following outlines the content of the five books in the series.

**Book 1**  **Managing Reality: Introduction to the Engineering and Construction Contract**

**Chapter 1**  **Introduction to the Engineering and Construction Contract, concepts and terminology**

This chapter looks at:

- An introduction to the ECC
- An identification of some of the differences between the ECC and other contracts
- An outline of the key features of the ECC
- Conventions of the ECC
- Concepts on which the ECC is based
- Terminology used in the ECC
- Terminology not used in the ECC
- How the ECC affects the way you work

**Chapter 2**  **Roles in the Engineering and Construction Contract**

This chapter describes the roles adopted in the ECC, including:

- How to designate a role
- Discussion of the roles described in the ECC
- Discussion of the project team
- How the ECC affects each of the roles

**Appendix 1**  **List of duties**

**Book 2**  **Managing Reality: Procuring an Engineering and Construction Contract**

**Chapter 1**  **Procurement**

This chapter looks at the concept of procurement and contracting strategies and discusses:

- Procurement and contract strategy
- What tender documents to include in an ECC invitation to tender
- How to draft and compile a contract using the ECC
- Procurement scenarios that an employer could face and how to approach them
- What are framework agreements and how they could incorporate the ECC
- What is partnering and how it can be used with the ECC

**Appendix 1**  **Assessing tenders**

**Appendix 2**  **ECC tender documentation**

**Chapter 2**  **Contract Options**

This chapter looks at the Contract Options available within the ECC:

- ECC main and secondary Options
- Priced contracts
- Target contracts
- Cost-reimbursable contracts
- Choosing a main Option
- Choosing a secondary Option

- Emphasises the importance of early dispute resolution to the successful outcome of a contract
- Considers the common sources of dispute
- Considers how the ECC has been designed to reduce the incidence of disputes
- Examines how the ECC provides for the resolution of disputes
- Looks at the implications for the dispute resolution process as a result of the new Housing Grants, Construction and Regeneration Act 1996
- Looks at ECC3 changes in relation to adjudication

This chapter describes the following:

- The compensation events contained within the ECC
- Procedure for administering compensation events
- Roles played by the two main parties to the contract in relation to compensation events

This chapter discusses aspects relating to the Schedule of Cost Components including:

- When the Schedule of Cost Components is used
- How the SCC interacts with the payment clauses
- Actual Cost and Defined Cost
- The Fee
- The components of cost included under the Schedule of Cost Components
- Contract Data part two

This chapter brings together all the aspects discussed in previous chapters in Books 1 to 4, which form part of the series of books on Managing Reality. This chapter provides the 'how to' part of the series. It introduces some example pro-formas for use on the contract.

For quick reference, this chapter may be read on its own. It does not, however, detail the reasons for carrying out the actions, or the clause numbers that should be referred to in order to verify the actions in accordance with the contract. These are described in detail in other chapters that form part of this series.

# List of figures

# List of tables

**Procuring an Engineering and Construction Contract**
ISBN 978-0-7277-5720-3

# Introduction

**General**

This series of books will provide the people who are actually using the Engineering and Construction Contract (ECC) in particular, and the New Engineering Contract (NEC) suite in general, practical guidance as to how to prepare and manage an ECC contract with confidence and knowledge of the effects of their actions on the Contract and the other parties.

Each book in the series addresses a different area of the management of an ECC contract.

- Book One – Managing Reality: Introduction to the Engineering and Construction Contract.
- Book Two – Managing Reality: Procuring an Engineering and Construction Contract.
- Book Three – Managing Reality: Managing the Contract.
- Book Four – Managing Reality: Managing Change.
- Book Five – Managing Reality: Managing Procedures.

- *Book One (Managing Reality: Introduction to the Engineering and Construction Contract)* is for those who are considering using the ECC but need further information, or those who are already using the ECC but need further insight into its rationale. It therefore focuses on the fundamental cultural changes and mind-shift that are required to successfully manage the practicalities of the ECC in use.

- *Book Two (Managing Reality: Procuring an Engineering and Construction Contract)* is for those who need to know how to procure an ECC contract. It covers in practical detail the invitations to tender, evaluation of submissions, which option to select, how to complete the Contract Data and how to prepare the Works Information. The use of this guidance is appropriate for employers, contractors (including subcontractors) and construction professionals generally.

- *Book Three (Managing Reality: Managing the Contract)* is essentially for those who use the contract on a daily basis, covering the detail of practical management such as paying the contractor, reviewing the programme, ensuring the quality of the works, and dispute resolution. Both first-time and experienced practitioners will benefit from this book.

- *Book Four (Managing Reality: Managing Change)* is for those who are managing change under the contract; whether for the employer or the contractor (or subcontractor), the management of change is often a major challenge whatever the form of contract. The ECC deals with change in a different way to other more traditional forms. This book sets out the steps to efficiently and effectively manage change, bridging the gap between theory and practice.

- *Book Five (Managing Reality: Managing Procedures)* gives step-by-step guidance on how to apply the most commonly used procedures, detailing the actions needed by all parties to comply with the contract. Anyone administering the contract will benefit from this book.

**Background**

The ECC is the first of what could be termed a 'modern contract' in that it seeks to holistically align the setting up of a contract to match business needs as opposed to writing a contract that merely administers construction events.

The whole ethos of the ECC, or indeed the NEC suite generally, is one of simplicity of language and clarity of requirement. It is important that the roles and responsibilities are equally clear in definition and ownership.

When looking at the ECC for the first time it is very easy to believe that it is relatively straightforward and simple. However, this apparent simplicity belies the need for the people involved to think about their project and their role, and how the ECC can deliver their particular contract strategy.

The ECC provides a structured flexible framework for setting up an appropriate form of contract whatever the selected procurement route. The fundamental requirements are as follows.

- The Works Information – quality and completeness – what are you asking the Contractor to do?
- The Site Information – what are the site conditions the Contractor will find?
- The Contract Data – key objectives for completion, for example, start date, completion date, programme – when do you want it completed?

The details contained in the series of books will underline the relevance and importance of the above three fundamental requirements.

**The structure of the books**

Each chapter starts with a synopsis of what is included in that chapter. Throughout the book there are shaded 'practical tip' boxes that immediately point the user towards important reminders for using the ECC (see example below).

> Clarity and completeness of the Works Information is fundamental.

There are also unshaded boxes that contain examples to illustrate the text (see example below).

> Imagine a situation in which the *Supervisor* notifies the *Contractor* that the reinstatement of carriageways on a utility diversion project is not to the highway authority's usual standards. However, the Works Information is silent about the reinstatement.
>
> Although it is not to the authority's usual standard, it is **not** a Defect because the test of a Defect is non-conformance with the Works Information. In this situation, if the *works* need to be redone to meet the authority's requirements, the *Contractor* is entitled to a compensation event because the new requirements are a change to the Works Information.

Other diagrams and tables are designed to maintain interest and provide another medium of explanation. There are also standard forms for use in the administration and management of the contract, together with examples.

Throughout the books, the following terms have been used in a specific way.

- NEC is the abbreviation for the suite of New Engineering Contracts and it is not the name of any single contract.
- ECC is the abbreviation for the contract in the NEC suite called the Engineering and Construction Contract.

The NEC suite currently comprises the

- Engineering and Construction Contract
- Engineering and Construction Subcontract
- Engineering and Construction Short Contract
- Engineering and Construction Short Subcontract
- Professional Services Contract
- Adjudicator's Contract
- Term Service Contract
- Term Services Short Contract
- Supply Contract
- Supply Short Contract
- Framework Contract.

**Procuring an Engineering and Construction Contract**
ISBN 978-0-7277-5720-3

# Chapter 1
# Procurement

**Synopsis**

This chapter looks at the concept of procurement and contracting strategies and discusses

- procurement and contract strategy
- what tender documents to include in an ECC invitation to tender
- how to draft and compile a contract using the ECC
- procurement scenarios that an employer could face and how to approach them
- what are framework agreements and how they could incorporate the ECC
- what is partnering and how it can be used with the ECC.

## 1.1 What is procurement?

Procurement can be defined as obtaining by purchase, lease or other legal means, plant, equipment, materials, services and works required by an organisation.

The process of procurement commences when a need (the accurate definition of this business need is essential prior to attempting to undertake any form of procurement process) has been established and ends when the project has been delivered. The business need must reflect the full-life value of the project, that is, what the project must deliver over a defined period. It should be recognised that in many instances the project goes beyond the physical works but also extends into the operational phase of a project.

Procurement can be a lengthy and complex process and the potential for waste and error is high. The procurement process usually involves a range of different stakeholders and people's expectations of the process need to be carefully managed in terms of time-scale issues and the fact that stakeholder needs will vary, as well as the usual needs/wants dilemma. The more effort and thought that is invested in the procurement process, the lower the potential for disputes during the contract period.

While it is often tempting to try to procure a solution quickly, it is important that sufficient time is allowed for effective procurement to take place. It is important that all factors should be considered, encompassing the technical, legal and commercial aspects of the project and the accompanying risks. Narrowing selection criteria to consider price and the end-product may only distort, or ultimately prevent, the achievement of the original objectives of the procurement exercise.

It is advisable or even essential that, wherever possible, all necessary commercial and contractual discussions have been executed prior to work commencing. Failure to do so will mean that both parties are working at risk and could result in disagreements, which are difficult to resolve, as the intended and actual basis of the proposed contract may be open to different interpretation by each party (or the basis of the original agreement could even be forgotten).

> Procurement involves not only commercial and contractual matters, but also technical, health and safety, environmental, security and all other aspects of the project.

When considering the procurement of a solution to meet a business need, employers should use all of their knowledge on not only the technical aspects of the product or works but also on the commercial aspects. The following is a list of the types of items that should be considered when planning a procurement exercise

- operational requirements/end-user consultation
- the employer's requirements/specification, whether functional, performance or technical (quality, fitness for purpose, minimum performance standards)
- the time frame (programme)
- the lifetime or 'whole-life' cost of the project (life-cycle costing/costs in use)
- possible tender list
- pricing
- required level of insurance cover
- environmental considerations
- health and safety matters
- security
- risk
- project inhibitors (e.g. site constraints/site limitations/availability)
- impact on business (e.g. disruption to business activities during works).

## 1.2 Value for money

The concept of value for money is not the same as accepting the lowest priced offer. A value-for-money approach is a move away from the policy of accepting lowest-price tenders, in the recognition that the value of the project to the organisation is far more than just the out-turn cost of the project, and includes the capital and revenue costs over the estimated life of the project. It is well known in the industry that lowest-priced tenders do not necessarily equate to best value for money and in fact may result in larger and more frequent management problems on the project resulting from claims. Furthermore, the appointment of a contractor on a value-for-money basis is likely to reduce the risk of project failure.

> Value for money includes non-price items as well as price.

Value for money means assessing the optimum balance of whole-life cost cycle, construction cost, time and quality to meet the client's requirements. Some aspects of tenderer's proposals which might be considered (beyond the tender price itself) include:

- acquisition costs
- procurement and head office costs
- maintenance costs
- management, operating and disposal costs
- the out-turn quality and the time taken to complete the project
- information about the tenderer's organisation, for example organisational structure
- tenderer's supply chain management
- tenderer's relevant previous experience
- how the tenderer intends to manage the resources on site, the quality and health and safety on site
- how the tenderer would respond to a described situation
- corporate social responsibility
- and, of course, the health and safety plan, the programme, method statement, design proposal, financial robustness, environmental policy, list of subcontractors, and other contract-specific items.

The tender evaluation process should recognise not only the tender price of the tenderer's proposals, but the other aspects of the proposal which are likely to contribute to the achievement of overall value for money (see above list).

Communicating a value-for-money approach to tenderers during the tendering process may also discourage prospective tenderers from attempting to 'buy' jobs (in the hope of later turning a profit through the issuing of claims).

## 1.3 The procurement process

The procurement process for works projects (see Figure 1.1) typically involves the following steps.

1 **Define need**
Identifying and clearly defining the business need.

**Figure 1.1** Procurement process

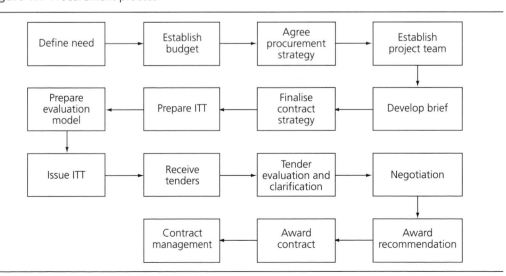

2 **Establish budget**
Obtaining the necessary funding for the planned expenditure.

3 **Agree procurement strategy**
Discuss with relevant parties and agree the procurement strategy for the overall project.

4 **Establish project team**
- *Appoint consultants*
  Procuring the services of external consultants for feasibility/design (if required) (which involves a separate procurement exercise).
- *Establish contractor input*
  Commencing discussion with contractors (if required) (using an appointment) for initial comments on design, buildability, etc.

5 **Develop requirement (brief)**
Developing the scope of work (or employer's requirements if a design-and-build approach is to be adopted). An initial risk assessment would also be undertaken at this time.

6 **Finalise contract strategy**
Discussing the contract strategy, taking into account the wider procurement strategy, the completeness/status of the scope of work and the requirements of the employer specific to the contract in terms of delay damages, retention and so on.

7 **Prepare tender documentation (invitation to tender)**
Producing tender documentation and identifying the tender list (prospective tenderers), drafting contract documentation.

8 **Prepare evaluation model**
Developing the tender evaluation model and considering the necessary range of value-for-money criteria. This model should contain the commercial : non-commercial weighting. Each descriptor under each of the commercial and non-commercial aspects should also be weighted for scoring purposes.

9 **Issue ITT**
Issuing the invitation to tender letter with the accompanying tender documentation (ensuring that the conditions of tendering are separated from what will eventually become contract documentation).

10 **Receive tenders**
Receiving the tenders and formally logging them.

11 **Tender evaluation and clarification**
- Evaluating tenders in terms of commercial and non-commercial aspects (e.g. technical, quality). Ideally, the technical evaluators do not receive priced tenders. For some traditional contracts, the commercial manager may be required to check the prices submitted by the tenderers and provide an evaluation report.
- Clarification of elements may be formally sought if required (and before evaluation can be finalised).
- Establish shortlist or identify preferred bidder if this is clear.

12 **Negotiation**
Carry out negotiation (if required) with shortlist or preferred bidder.

13 **Award recommendation**
Compiling the contract documentation using the tender documentation issued at invitation to tender stage and adjusting it, taking into account any proposals and changes and at the negotiation stage. Seeking formal authority to award a contract based upon this final set of contract documentation.

14 **Award contract**
Issuing the contract documentation in duplicate original to the contractor for his signature and countersigning documents on their return. One fully executed original should then be returned to the contractor as his copy of the contract.

15 **Contract management**
The contract management process then takes over. This includes post-project evaluation and, ultimately, post-occupancy review (where applicable). Monitoring and reviewing the process (for procurement and project delivery) to facilitate continuous improvement.

Figure 1.2 The procurement process timeline

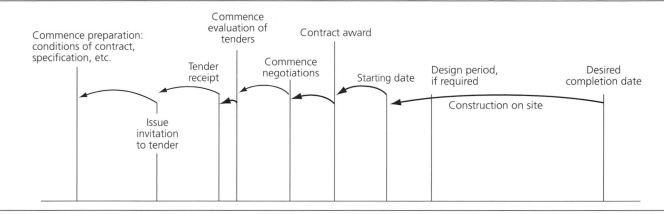

## 1.4 Procurement time cycle

It is important whenever possible to leave sufficient time during the procurement process to ensure that the required procedures, process and policies can be followed. In some instances, execution of the full process will not be possible, for example, emergency or safety-related works. In addition, it is advisable to establish an accelerated procurement process which can be adopted where there is a critical operational need.

Generally, the time cycle for the procurement process should be measured back from the intended date for completion or operation of the project to determine the necessary starting time for the procurement process (see Figure 1.2).

Clearly, reasonable estimates as to the appropriate time allowances allocated for each stage of the procurement process activities need to be made. If the procurement is subject to the EU procurement rules, then the time periods for certain activities are dictated (e.g. where the value of a works project is greater than its EU threshold, the choice of procedure dictates the time that must be given to register interest or return the tender. The EU regulations apply to public bodies and utilities only) and any key dates (the key dates might be the date for prospective tenderers registering interest and the final date for tender return) should be incorporated into the procurement time frame.

Allowing insufficient time for the procurement process will inevitably result in inadequate time being left for concluding the process – for example, evaluation and contract drafting – and could jeopardise the quality of the contract that is achieved.

> Leave adequate time for procurement in the project programme.

It is often early on, in the procurement stages of the project, where the greatest benefits can be derived. These benefits might include obtaining better value for money or improved quality, reducing time, or ensuring that a particularly important issue is properly addressed. Good preparation can greatly enhance performance during and on completion of the project – like many things in life, the greater the effort expended initially, the greater the rewards.

## 1.5 Procurement strategy

An important part of any procurement strategy is recognising the parameters under which the strategy operates. Not only will there be internal policies and procedures to follow regarding the issuing of tender documentation and ensuring auditable processes, but there are legal, statutory and regulatory requirements to follow as well.

### 1.5.1 What is procurement strategy?

A procurement strategy involves looking at the longer-term, wider procurement picture (such as long-term agreements) for an organisation, rather than merely considering the current contract (the subject of a procurement strategy cannot be fully discussed here and is mentioned only to emphasise that clients should consider a view that is not confined to the on-site construction time of one contract). A contract or series of contracts may be stepping-stones to the achievement of a broader procurement or business objective.

A procurement strategy involves considering:

1 Whether the contract is just one of many that make up a larger programme of work.
2 How the project fits in with overall business activity, that is:
   - how many other projects are being constructed on the same site, affecting interfaces, resources and cost analysis if site set-up is shared,
   - how many other projects are ongoing or due to commence in the organisation and which may require use of the same resources.
3 How the procurement activity required for the project fits into the organisation's overall procurement activity plan:
   - for example, the organisation might already have arrangements in place with specialist suppliers or subcontractors or might consider procuring separately the concrete required for a large building project and free-issuing it to the contractor, if it considers that this would prove more economic than having the contractor source it.

## 1.5.2 Procurement route

A narrowed view of procurement strategy for a contract could include considering the procurement route that will be adopted for the contract, that is, how is it best to procure the contract?

- Using any existing framework contractors (see section 1.9 below).
- Utilising a two-stage tender process.
- How early in the procurement process to involve the contractor, that is, before the design stage, during the design stage or only at build stage?

## 1.6 Contract strategy

The contract strategy contains the details chosen for the particular project. This should take place after the procurement strategy has been decided upon and implemented since the overall procurement strategy affects the way in which the contract is procured.

> - Procurement strategy considers the wider impact on the business.
> - The procurement route for a contract includes considering how to procure that contract.
> - Contract strategy includes deciding what the contract will include (liquidated and ascertained (L&A) damages, retention, bonus for early completion, etc.).

The contract strategy involves deciding how the conditions of contract used help you to achieve the objectives of the project.

- Do you need retention?
- Do you need delay damages?
- Do you want to take over parts of the work as they are completed?
- How do you want to pay the contractor?

You may also want to consider the following as part of the contract strategy.

- With whom do you wish design responsibility to lie (design team, contractor)?
- Who will take responsibility for project delivery?

Most of these questions are answered through the choices made in the Contract Data (see Chapter 3 on the completion of the Contract Data).

## 1.7 Public sector procurement

In the background of any procurement in the public sector lie the EU procurement rules (currently applicable to public authorities and utilities only). Any procurement should follow the award procedure referred to in the notice. The award criteria should be included in the invitation to tender, and a notice confirming the successful tenderer should be lodged upon completion of the procedure.

> The EU procurement regulations affect only public bodies and utilities.

Procurement in the public sector is tightly regulated and auditability and public accountability must be ensured at all times. The procurement process is often more detailed and involved than that of the private sector.

The procurement process in the public sector must ensure

- a clear audit trail
- the ability to demonstrate clearly that value for money has been achieved
- that compliance with all applicable regulations/statutory and legal requirements has been met at all times.

## 1.8 Generic good principles for effective procurement

The principles of procurement for any project are the same. This section is therefore aimed at providing a reminder and overview of some of the golden rules to apply when preparing documentation for procurement purposes, so that some of the pitfalls that befall procurement activities can be avoided.

The tender documents should be originated specifically for each project. Any temptation to copy or amend previous similar projects should be resisted, as this may lead to the occurrence of anomalies and inconsistencies resulting from some or all of the following.

- Using previous specifications that have not been updated or that include irrelevant information.
- Using previous specifications that have been amended to suit the previous contract. (It may be that previous conditions of contract or paragraphs in the specification were introduced by a project manager who had been affected by their lack in the past. These conditions can be onerous and draconian and may simply not be suitable to all contracts or all contractors.)
- Using the same conditions of contract as were used in previous contracts, without considering how appropriate they are to the current project.
- The inclusion of every possible document in the filing cabinet to 'cover' you and make sure that nothing has been left out, rather than considering the appropriateness of the document. This will almost certainly lead to a confused contract with duplication or conflict between the documents.
- Including conditions of tendering and tender terminology throughout the document (rather than confining them to a document for conditions of tendering; tender terminology should not be included in specifications, in the employer's requirements or in the conditions of contract); for example, that the tenderer is to include a programme with his tender, that the tenderer is to have taken all information about the site into account when tendering, that the tenderer is to have included for all items in his pricing.

### 1.8.1 Letters of award

It is the practice of many employers to issue a letter of award to the successful tenderer. This letter states that the tenderer's offer has been accepted, and that the contract 'includes the following'. The list that ensues includes the tender documentation, the tenderer's offer, any subsequent offers, and communications and minutes of meetings in which anything relevant to the contract appears.

Because there are elements of the tender documentation, such as the rules of tendering, that do not form part of the contract document, it is inadequate simply to state in the letter of award that the contract comprises the tender documentation, the tenderer's offer, and all communications up to a particular date. This potentially introduces conflicts, ambiguities and a source of disputes that could be avoided very easily. Conflicts between the tender documents and the tenderer's offer may result, and confusion arises as to which prevails. For this reason, it is advisable to include an order of precedence in the letter of acceptance. However, if a dispute were to occur, it might still be extraordinarily difficult to ascertain, among all the paper that makes up the contract, what the contract actually says and in whose favour the dispute may lie. Even if an order of precedence is stated, the *Contractor* might have a good reason for stating that his tender was not made on the premise argued by the *Employer*.

The more work that is invested prior to contract award, the more aware the *Employer* should be of exactly what he is contracting for.

A letter of award should simply state that the *Contractor* has been awarded a contract. The entire contract should then be attached and should include all the documents that comprise the contract. The 'entire document' in this sense does not mean a copy of the tender and any other communications, but rather means a copy of the documents that make up the contract, as amended for any changes made in the tender and other communications. The tender documentation and the tenderer's offer **in their entirety** are not part of the contract, although some of the documents within them will form part of the contract, including

- the form of contract/Articles of agreement, completed and available for signing
- the conditions of contract, amended as agreed between the parties
- the final prices as negotiated between the parties
- the definitive specification (or employer's requirements and contractor's Proposal for a design-and-build contract).

In this way, both parties know what the contract comprises and this sets the basis of the entire contracting relationship, including the obligations of the parties to one another.

**1.8.2 Letters of intent**

Letters of intent are complicated and potentially leave one of the parties unprotected. Issuing a letter of intent should be avoided if at all possible. The employer should never be in a situation where in order for the work to start, a letter of intent has to be issued because the contract is not in place. Procurement should take place with an efficiency that allows the contract to be in place before work is required to start.

**1.9 Long-term agreements**

**1.9.1 A manageable supply base**

Many organisations consider their spend on a project-by-project basis. Separate teams are arranged for each project and there is continuity over the spend only to the extent that the same bidders might continually win work. Figure 1.3 represents this procurement strategy.

The organisation's ability to drive strategic improvements is facilitated by moving away from a project-by-project focus to a business and programme focus. See Figure 1.4.

Organisations with a business and programme focus take project delivery seriously and do not consider cost, time and quality drivers in isolation. Provided the organisation can think strategically, it has the opportunity to manage its supply chain more effectively. By taking a programme approach to the business spend, the group of projects is considered together and the supply chain can be consolidated to facilitate working more effectively with a reduced and more manageable group of suppliers. The most obvious way to do this and

**Figure 1.3** Project-by-project focus

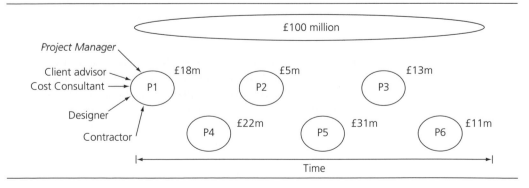

**Figure 1.4** Business/programme focus

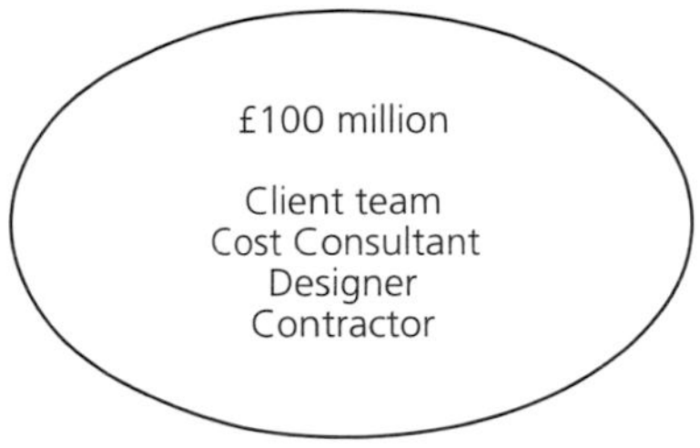

> £100 million
>
> Client team
> Cost Consultant
> Designer
> Contractor

to prove value for money is to put in place long-term agreements. In this way, the team can work together to achieve the targets.

### 1.9.2 Continuous improvement

Procurement is a costly operation that is critical to the overall success of a business. A key driver for businesses, whether in the private or public sector, is to improve the quality of the product or service and to drive out waste, unnecessary costs in machinery, materials, labour, procedures and practices. In terms of construction, this has seen a focus to drive out waste in the procurement of works, goods and services. A useful vehicle through which continuous improvement of this nature can be achieved is the long-term agreement.

> Procurement is a critical business activity.

### 1.9.3 Long-term agreements

A long-term agreement is where the parties agree to focus on long-term business improvements for the employer's organisation. Examples of the benefits sought are

- reduced construction period (increased speed to market)
- enhanced safety environment/culture
- assurance of supply (including availability when required and of the standard required)
- improved reliability (project delivery on time, to cost and quality)
- reduced costs through familiarity, increased efficiency of delivery
- better-quality solutions achieved over time
- efficiency of working (process improvements, faster learning, increased productivity over time).

The parties are able to do this through working in partnership over a prolonged period. This also improves contractor understanding of the employer's business, the procedure and the types of solutions required. By working under a long-term agreement, continuous improvement targets can be set against a range of criteria and a sustained effort made to meet these. Examples of the potential benefits which might be realised (and for which targets could be set) include the following.

1 Workload security, for example:
   - continuity of workload/planned programmes of work (security of work in a recession) and
   - relationship continuity and development, for example:
     - continuity of people/skills/competencies/expertise
     - economies of scale
     - improved product and services (e.g. investment in new technology/techniques)
     - consistency of approach
     - securing skills and scarce resources (e.g. skill shortages in an overheating marketplace either generally or for specialist skills).
2 Enhanced relationship with the contractor, leading to a partnering attitude and a unified approach to the work being done.
3 Contribution of ideas and designs that would assist the employer in his business.
4 Marketplace/economies of scale.

**Figure 1.5** Economic cycle

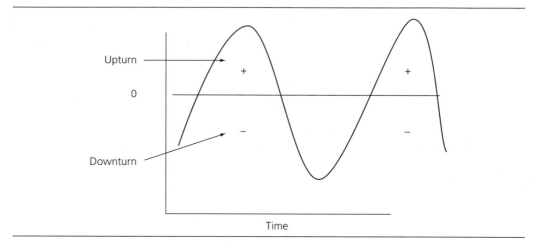

**Figure 1.6** Features of the economic cycle

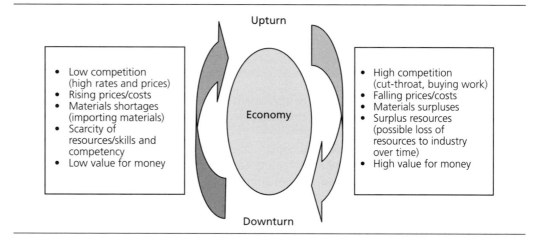

5 Less wasteful tendering.
6 More trusting relationships.

The benefits sought obviously depend on many factors, including market conditions, market trends, business objectives, etc. Figures 1.5 and 1.6 show some of the influencing factors. There is a distinct difference between best practice procurement and the optimisation of procurement (economics, principles, climate).

Long-term agreements should take a flexible view that is not dependent on the climate – that is, that does not take advantage of recessions in the economy (see Figure 1.7).

**Figure 1.7** Vision versus view

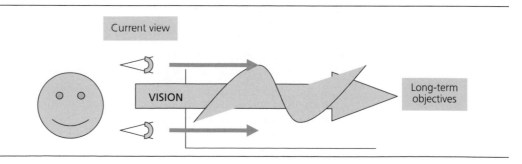

This could involve

- not including liquidated and ascertained (L&A) damages into the contracts, since a long-term view is being taken
- not including retention, since retention indicates a short-term view
- encouraging innovation in the marketplace even when times are tough
- considering the benefits of an upturn and a downturn
- seeing through to the long term – vision versus view.

## 1.10 How does the ECC contribute to effective 'partnership'?

The ECC inherently accommodates the latest vogues for partnering, long-term agreements, alliances, frameworks and the like. The procedures that the ECC uses can be seen to lie in a similar or indeed 'back-to-back' fashion with 'Partnering'. Some of these can be categorised as follows.

### 1.10.1 Contract administration

- Addresses problems as and when they occur.
- Effective communication tool awakens responsibility. The ECC encourages people to communicate; it proposes sanctions upon *Contractors* or *Project Managers* who fail to perform diligently.
- Smoothes the way to focusing on real issues and progressing the job.
- Compensation events are a risk-levelling device not allowing hidden costs and avoiding tail-end claims.
- Quotation procedure allows flexibility – that is, full or shorter method at the discretion of the contract parties.

### 1.10.2 Dispute resolution

- Adjudication as a first step in lieu of arbitration or litigation.
- Joint appointment of *Adjudicator*.

### 1.10.3 *Employer* contribution

- Through the *Project Manager*, decision making is not left entirely to designers or architects.
- As the contract cannot be left in the 'bottom drawer', the *Employer* is forced to address real issues and take an active involvement in his project's maturation.

The contract recognises that the *Employer*, having expressed a desire to enter into the contract, has to maintain responsibility under it and so via his *Project Manager* must take an active role in its performance.

### 1.10.4 Multiplicity of use

As a set of interlocking documents, the ECC allows uniformity and consistency across a full spectrum of contract types, for example, building, civils, professional services. This ability to interlock has a crucial benefit for 'partnerships' and is explored in a little greater depth in the next section.

### 1.10.5 Trust

By the very fact that a Client is prepared to adopt the ECC, he is sending out a clear signal to prospective Contractors and Consultants that his attitude has changed. It is a statement or declaration of intent to which we must all respond.

> The ECC embodies many aspects of partnering.

### 1.10.6 Suite/family of contracts

The ECC offers the potential to encompass the majority of procurement and project routes through its main and secondary Options.

## 1.11 ECC tendering procedure

The ECC tendering procedure is represented diagrammatically in Figure 1.8 and discussed in the following subsections.

### 1.11.1 General

When issuing an invitation to tender letter and tender documentation, the most important thing to remember is that these documents will form the basis of the resulting contract. The tender documentation should not be incorporated into the contract by reference, but should be redrafted to become the contract. In other words, the tender documents

**Figure 1.8** The ECC tendering procedure

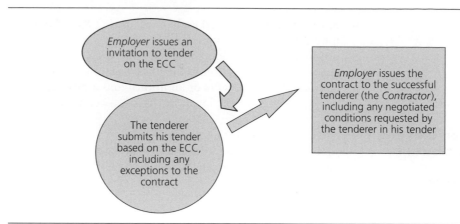

should not be listed in the contract award letter as part of the contract; they should be drafted in such a way that parts of them can be discarded and not form part of the contract.

Similarly, the tender should not be incorporated into the contract by reference. Those parts of it that amend the tender documentation should be extracted and the contract amended accordingly.

Many employers will consider a redraft far too much work. It may be that for smaller contracts, where the risks are lower, the work required for a redraft could exceed the benefits in terms of consistency and lack of ambiguity. It might be considered worth the effort of getting into a routine of general amendment, however. In whatever manner the employer chooses to issue the contract – whether through redraft or whether by total inclusion of all tender and post-tender documentation – the employer should ensure that ambiguities and inconsistencies are removed from the document in order to decrease the potential for conflict.

> The tender documentation should not be incorporated into the contract by reference, but should be redrafted to become the contract.

**1.11.2 *Employer's* requirements**

The tenderer will submit a compliant bid for the *works* based on the tender documentation (Contract Data, Works and Site Information, etc.) sent to him by the *Employer*. The original tender documentation may be changed by the *Employer* during the tendering period by the issuing of amending letters. This will form the basis of the information upon which the tenderer's price is based.

These requirements will form the basis of the *Contractor's* obligations under the contract. Care should be taken to ensure that any amendments or qualifications made by the *Contractor* as part of his submission are clearly identified. In these instances the *Contractor* will have submitted a non-compliant bid.

**1.11.3 *Conditions of* contract**

Any amendments to the *conditions of contract* suggested by the tenderer and agreed to by the *Employer* should be incorporated into the contract by redraft rather than by reference. A reference to a letter or the tenderer's bid, where the redrafted condition appears, could cause confusion and conflict, particularly if the amendment is not in an obvious and easily accessible place. If the *Employer* accepts the change, it is more effective to include the change in a noticeable place.

**1.12 From tendering to contract**

It can be seen from the example below that the invitation to tender (ITT) forms the basis of the contract. Although the ITT may comprise different elements, in general, the items listed below will be included.

The **tender** documents comprise the following.

1  PART A
   - Letter of invitation
   - Conditions of tendering
2  PART B
   - Tender submission document
3  PART C
   - Form of contract/Articles of agreement
   - Contract Data
   - Contract Prices
   - Works Information
   - Site Information

The **contract** comprises the following.

- Award Letter.
- Form of contract/Articles of agreement.
- Contract Data parts one and two.
- Contract Prices.
- Works Information.
- Site Information.

The following gives a brief explanation of Parts A to C of the tender documents.

1  PART A
   The part of the tender documents that describes the conditions of tendering and the instructions to tenderers. This part is not included in the resulting contract.
2  PART B
   The part of the tender documents that relates to how the tenderer is to respond; that is, the form of tender and the schedules he is to complete such as CVs, project organisation charts, health and safety policy, a programme and method statement (statement of how the *Contractor* plans to do the work). This part is not included in the contract, although the response submitted by the tenderer incorporating this part could be included in the contract.
3  PART C
   The part of the tender documents that forms the basis of the contract; that is, the *conditions of contract*, the *Employer*'s requirements and the agreement that will be signed between the *Employer* and the successful tenderer. There should be no reference in this part to the 'tender' or the 'tenderer'.

The items that are specifically related to tendering (Parts A and B) should be kept in one part of the tender documents. These Parts A and B do not form part of the eventual contract and can effectively be detached and removed when pulling the contract together.

There should therefore be no reference to the tender or tenderers in the employer's requirements or any other document that will form part of the contract (Part C). There should only be reference to the *Contractor*, the person who will ultimately be responsible for carrying out the work.

### 1.12.1 Structure of an ECC invitation to tender from the *Employer*

A possible structure for ECC tender documentation is as follows (see Appendix 1 to this chapter for an example of the ITT format).

- Letter of invitation.
- Conditions of Tendering.
- Tender submission documents (i.e. what the tenderer is to include with his tender, for example, a statement of how the *Contractor* plans to do the work).
- Form of contract/Articles of agreement.

- Contract Data part one (completed).
- Contract Data part two (left blank – to be completed by the tenderer).
- Contract prices (blank form to be completed by the tenderer).
- Works Information and Site Information.

### 1.12.2 Structure of an ECC tender from the *Contractor*

A possible structure for a tender based on the ECC is as follows.

- Letter from the tenderer.
- Tender submission documents completed.
- Contract Data part two (completed).
- Contract prices (completed).
- Compliance with the Works Information or the tenderer's proposal where it is a design-and-build or a performance specification.

### 1.12.3 Structure of an ECC contract

A suggested structure for a contract based on the ECC is as follows.

- Form of contract/Articles of agreement.
- Contract Data parts one and two.
- Contract prices (*activity schedule* or *bill of quantities*).
- Works Information.
- Site Information.

This is the simplest way to present a contract using the ECC. The form of contract/Articles of agreement is a written document that facilitates the signature of both parties and states the Parties' principal obligations – that is, that the *Employer* will pay the *Contractor* for the work done in accordance with the contract; and that the *Contractor* will carry out the *works* in accordance with the contract.

The form of contract/Articles of agreement is a mechanism for contract execution. It is part of a bound contract document where the rest of the contract is exactly as agreed between the Parties; that is, the documents have been amended to reflect any changes made during the tendering process.

Where the *Employer* does not choose this mechanism, he is likely to use a contract award letter accepting the tender sum as stated in the form of tender and listing the documents that are included in the contract, such as

- the invitation to tender
- the tender
- the outcome of any post-tender negotiation.

As stated above, however, this particular method may introduce potential disputes unnecessarily and too early in the project life cycle.

### 1.13 Procurement scenarios
### 1.13.1 Introduction

The following scenarios are examples of some of the procurement routes available to employers.

1. The *Employer* has fully designed the required *works* and requires a lump sum price from the *Contractor* for the construction of the *works*.
2. The *Employer* has provided a performance specification to the *Contractor* and requires him to design and construct the *works* on a fixed sum basis.
3. The *Employer* requires the *Contractor* to design and construct the *works* and to participate in a target cost contract.
4. The *Employer* requires immediate and urgent *works* that are not designed.
5. The *Employer* requires a feasibility study to be completed, a working up of the specification, and the construction of the *works*.

The choice of ECC main Option is based on the completeness of the Works Information, as well as on the informed decision of the *Employer*. In general, there is no need to use anything other than an Option A contract, where the design is fully developed by the *Employer*'s designers or the *Contractor* and the work is not urgent.

Where the work is designed by the *Employer*, the completeness of the design (and possibly other factors, such as the incentivisation of the *Contractor*, past relationships (frameworking) and the current climate and situation financially, politically, etc.) will determine whether Option A/B or C/D is used. The use of Option C or D is a choice made by the *Employer* according to whether the scope can support a fixed price or not. In other words, where the Works Information is defined sufficiently to support an Option A contract, then Option A would be the payment option that would minimise the *Employer*'s risk. Despite this robustness of the Works Information, the *Employer* could still choose Option C if he wishes to amend the risk and flexibility structure of the contract and embark on a target cost contract.

Target cost contracts are often chosen for the wrong reasons or for inappropriate situations. Target cost contracts do not always provide an incentive to the *Contractor*, and the *Employer* is often left to pay far more than budgeted because of the target cost mechanism. Similarly, in situations where the design is fixed, whether by the *Employer* or the *Contractor*, a fixed-price contract is often more appropriate as a method of payment. Cost-reimbursable contracts are useful where the design is uncertain, and where a target cost contract would result in burdensome administration. In general, if the *Employer* is uncertain of the scope of his own design, then there will be additional costs, and a target cost mechanism may not reduce those costs.

## 1.13.2 Scenario 1: Works designed, fixed price

Where the *Employer* has previously engaged a designer to prepare the *Employer*'s requirements for a contract, and now requires a fixed price for providing the works, there are few steps involved in the procurement cycle.

Where the *works* are designed fully, the most appropriate payment route to adopt is that of a priced contract. This could be achieved using the ECC as Option A or Option B. Option B provides for the use of a *bill of quantities*, where a remeasurement route is chosen by the *Employer*. Option A uses an *activity schedule* (milestone payment schedule) as the tool for payment to the *Contractor* and is therefore a lump sum form of contract.

The steps in the procurement cycle are as follows.

1 Develop the *Employer*'s requirements through to a completed design (could be an internal appointment).
2 Check that the *Employer*'s requirements integrate with the *Employer*'s own special requirements.
3 Draft contract and tender documentation using Option A or Option B (could be an internal appointment).
4 Draft method of evaluation of tender.
5 Issue invitation to tender letter and accompanying tender documentation.
6 Evaluate the returned tenders on the basis of the Prices and the Schedule of Cost Components.
7 Commercially and contractually assess the tenders based on the method of evaluation.
8 Place the contract (in accordance with the award criteria included in the ITT or in any *Official Journal of the European Union* (OJEU) notice for public sector and utilities procurement).

## 1.13.3 Scenario 2: Works to be designed by *Contractor*, fixed price

Where the *works* are to be designed by the *Contractor*, a fixed-price contract is still appropriate; however, the *Employer* might prefer to use a target cost contract for other reasons. The procurement process is likely to be a one-stage or two-stage process in order to ensure that the most appropriate contractor against certain predetermined criteria is chosen. Whatever procurement route is chosen, the route should be spelled out in an EU notice (where this is required in accordance with the financial thresholds for a works contract for public sector bodies or utilities) and adhered to at all times.

A two-stage route could be chosen where there is insufficient time to develop the design in the tender period or where the *Employer* does not want the tenderers to bear the costs inherent in each of them developing a design. It should be noted that this route could result in greater *Employer* risk where a preferred contractor is chosen to develop the design. This is because the competitive tendering aspect is lost, and therefore the criteria for determining the best design are undermined. In addition to the increased commercial risk, case law does not support awarding a contract on percentages or non-lump sum prices that cannot be competitively compared effectively; therefore, the legal and commercial validity of a two-stage process may be inadequate.

### 1.13.3.1 One-stage process

A one-stage process might be used in conjunction with a value engineering process where the scope of work is partly uncertain. The process is based on a robust set of *Employer*'s requirements from which it is possible to tender a fixed price.

The steps in the procurement cycle are as follows.

1  Develop the *Employer*'s requirements.
2  Draft contract and tender documentation using chosen Option.
3  Draft method of evaluation.
4  Issue ITT letter and accompanying tender documentation for a design-and-build contract.
5  Evaluate the tenders on the basis of the Prices and the Schedule of Cost Components as determined by the evaluation model.
6  Determine technical acceptability of tender design.
7  Commercially and contractually assess the tenders.
8  Place the contract (in accordance with the award criteria included in the ITT or in any OJEU notice for public sector and utilities procurement).
9  Develop and commence the value engineering exercise to develop the design.
10  Notify and instruct a quotation for a compensation event to embrace the results of the value engineering exercise.

### 1.13.3.2 Two-stage process

Where the *works* are to be designed by the *Contractor*, a fixed price is generally the most appropriate method of payment throughout. This may be achieved using the ECC as Option A. Option A uses an *activity schedule*, or milestone payment schedule, as the tool for payment to the *Contractor*.

The steps in the procurement cycle are as follows.

1  Develop the *Employer*'s requirements.
2  Draft contract and tender documentation using chosen option.
3  Draft method of evaluation.
4  Issue ITT and accompanying tender documentation for a design-and-build contract.
5  Evaluate the tenders on the basis of the Prices and the Schedule of Cost Components or Shorter Schedule of Cost Components as determined by the evaluation model.
6  Determine technical acceptability of tender design.
7  Commercially and contractually assess the tenders.
8  Draft a Professional Services Contract for the development of the design by the *Contractor*, using Option A. The question of who owns the design should be answered using the contract conditions. This contract is finalised and executed to allow the *Contractor* to develop the design. The final output of this contract is a design and an offered sum for providing the *works* in accordance with the design. There are alternatives to this procurement route (see below).
9  A contract for the *works* is also drafted and executed but note that the reduced legal and commercial viability of such a contract (see the discussion in 1.13.3 scenario 2 above).

An alternative to engaging the *Contractor* on two contracts, one a Professional Services Contract (PSC) for the design of the *works*, and the other an Engineering and Construction Contract for the construction of the *works*, is to contract with the *Contractor* using only an ECC contract. There are advantages and disadvantages for both these methods as detailed below.

### 1.13.3.3 Using both a PSC and an ECC

| Advantages | Disadvantages |
|---|---|
| (1) The method of payment can be a fixed price throughout | (1) Slightly more work is involved in drafting two contracts instead of one; however, the work involved in shoehorning a works contract to fit design and construction could turn out to be considerably more |
| (2) The *Employer* has more power in negotiating the price for the works contract | |
| (3) The *Employer* can choose not to continue with the works contract | (2) The *Contractor* may be uncertain of receiving the construction contract, which may increase the price overall |
| (4) The works contract can still be drafted to include the design as belonging to the *Contractor*, thereby preserving the *Employer*'s contractual position | |
| (5) Using two contracts clearly sets out the Parties' contractual positions and allows the use of a more appropriate contract for the design of the *works* | |

### 1.13.3.4 Using an ECC only

| Advantages | Disadvantages |
|---|---|
| (1) Where a target cost contract is chosen for the design and construction, procurement is initially easier | (1) However, the target cost will have to be revised once the design is complete, and the new target either negotiated or devised using compensation events, resulting in a clumsy attempt to work backwards. In addition, any evaluation will have to be carefully thought through to adhere to case law that questions the viability of a two-stage process |
| (2) The negotiation of the price for the works contract could be intense since the *Employer* is already contracted with the *Contractor* and there is little room to manoeuvre | (2) Two different payment options could be chosen to cope with the two different aspects of the contract: design (Option A) and construct (Option E to be amended to Option A once phase 1 is complete). This results in negotiation and work to be completed part way into the contract |

> The ECC can be used for design-and-build contracts.

### 1.13.4 Scenario 3: Works to be designed by *Contractor*, target cost

Where the *works* are to be designed by the *Contractor*, a fixed price is generally the most appropriate method of payment throughout. There could be times, however, when, for reasons of his own, or due to the uncertainty of the *Employer*'s performance specification or the uncertainty of other conditions impacting on the works, the *Employer* might choose to use a target cost contract. This could be achieved using the ECC as Option C.

Option C uses an *activity schedule*, or milestone payment schedule, to ascertain the *Contractor*'s target cost – that is, what the *Contractor* thinks the *works* will cost if the described Works Information does not change. This target cost is amended throughout the contract when compensation events occur. The *Contractor* is paid his Defined Cost plus Fee throughout the period of the contract. The total of these payments is compared with the target cost at Completion and the difference is shared out between the *Employer* and the *Contractor* in a predetermined manner. The share mechanism is therefore intended to be an incentive to the *Contractor* to maximise his part of the share, but the *Employer*'s risk is increased.

In this case, the procurement decisions and actions are the same as for those of a fixed-price contract, except that Option A/B becomes Option C/D.

**1.13.5 Scenario 4: Works to start immediately without *Employer's* requirements**

Where work is required to start urgently with little idea of scope, an Option E contract is the most appropriate option.

If the contract for a public body or utility is large enough to be subject to EU Procurement Regulations, then competitive tendering is still required, increasing the time period before a contractor can commence. In some instances an accelerated procedure may be used. It could be that a call for competition could be dispensed with if it is strictly necessary for reasons of extreme urgency and the reasons that brought about the event were unforeseeable by the *Employer*.

Otherwise, there are the following steps in the procurement cycle.

- The contract is drafted with a brief description of requirements.
- The *Employer* holds discussions with various contractors to decide on the contractor to do the work, or a contractor is chosen through knowledge of previous works and industry experience.
- A contract is finalised through negotiation and evaluation.

**1.13.6 Scenario 5: Feasibility study, design and build**

It is not appropriate to use the ECC for all three stages of this procurement strategy. Some employers might choose to use the ECC including Option X5 (sectional completion); however, it is likely that consultants will be engaged for the feasibility study and possibly the design aspect, and therefore the use of one contract for two different contractors is not advised. Even if a contractor were to be engaged for all three items, by the very nature of a feasibility study, the design-and-build stage might not be undertaken if such a project is not deemed to be feasible.

It is more appropriate to engage a consultant using the PSC for the feasibility and design stages, and then to engage a contractor using the ECC after an invitation to tender (if required) using the design written by the consultant. An alternative is to engage a consultant using the PSC for the feasibility stage and then a contractor on the ECC for the design-and-build stage.

> It is not appropriate to use the ECC for feasibility, design and build using sectional Completion.

**1.13.7 Conclusion**

Note that these scenarios are simply that – scenarios. They do not form recommendations and they are not cast in stone. They are presented simply as a means to convey the various decisions that employers could face.

**Procuring an Engineering and Construction Contract**
ISBN 978-0-7277-5720-3

ICE Publishing: All rights reserved
doi: 10.1680/pecc.57203.023

# Appendix 1
# Assessing tenders

**Synopsis**

This appendix provides suggestions for assessing tenders, including

- what to ask for in the invitation to tender
- what should be submitted with the tender
- what the primary commercial and technical documents are
- what to consider in conducting the evaluation.

## A1.1 Introduction

Part of procurement is deciding how you will assess the tenders that are returned and what information you require and in what format you require it. All the normal aspects of assessing tenders exist with the ECC; there are however a few areas that need careful consideration.

Selecting the right contractor is very important. Public bodies and utilities are required to adhere to the EU Procurement Directives. Organisations that are not required to follow the EU Procurement Directives could still benefit from some of the guidelines and principles contained within the directives.

The assessment criteria should be based on your objectives for the project in both technical and commercial terms. These will in turn affect how you set up the tender documentation in the first place and what information you will call for at the time of tender.

## A1.2 Information to include in the invitation to tender (ITT)

The ITT should include all the information that you will need to assess tenders. This means asking for information to be included in the tender that will help you make your decision. It also means telling the tenderer how you intend to evaluate the tender, so that he is able to give you what you need to make your evaluation.

### A1.2.1 Information required from the tenderer

The ITT should list the information required from the tenderer. This information will assist the client in making his evaluation. Such information could include the following (see also section A2.5 of Appendix 2 to this chapter) and should be clearly separated into selection criteria and award criteria in line with recent case law (especially for EU procurements).

- Contract Data part two.
- A programme that shows how the tenderer intends to Provide the Works, including any statement of how the *Contractor* plans to do the work and resource statement.
- Organisational chart.
- Information about the people who will be working on the contract, such as curricula vitae.
- Pricing information.
- Environmental information and information on corporate social responsibility.
- Health and safety.
- Details about past projects.
- Partnering information.
- Investor in People.

### A1.2.2 Award criteria

In addition to listing the information that the tenderer is required to provide, the ITT should also include information on award criteria; that is, how the tenders are going to be evaluated.

Criteria weightings are a compulsory requirement of the EU Procurement Directives. In any case, it is suggested that a statement of award criteria is issued. An example is (see also section A2.4.5 of Appendix 2 to this chapter): most economically advantageous tender in accordance with the following criteria in descending order.

- Design innovation.
- Method statement.
- Programme.
- Tender price.
- Partnering approach and philosophy.
- Community benefits.

This imposes a discipline to think about the evaluation of the tenders before the invitation to tender letter and accompanying tender documentation are issued. It also facilitates the request for the information that will enable evaluation.

## A1.3 Evaluation
### A1.3.1 General

Some employers like to split the evaluation of tender submissions into technical and commercial parts. In such situations each of these elements of the tender will be delivered at the same time, but will be in separate packages to be reviewed on a technical and commercial

**Figure A1.1** Factors affecting the assessment of tenders

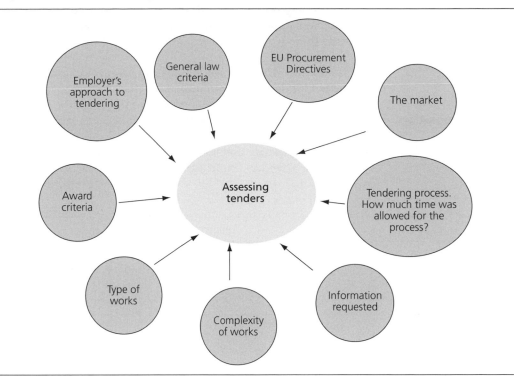

basis in isolation from each other. A reason for separating the technical and commercial elements of the tender is so that the technical reviewers are not influenced by price. Figure A1.1 shows some of the factors that influence the assessment of tenders.

**A1.3.2 Information submitted with tender**

The following is a list of the types of information, which may be requested and submitted with a tender (see also Appendix 2 to this chapter).

**Commercial**

1  Form of tender.
2  *Activity schedule* or *bill of quantities* (pricing document).
3  Contract Data part two.
4  Requested alternative prices or options.
5  Parent company guarantee/bonds (e.g. performance bond, advanced payment bond).
6  Design warranties for *Contractor* design.
7  Collateral warranties.
8  Insurance certificates.

**Technical** (Note that some of the following are selection criteria and some are award criteria; the two sets of criteria should be separated and addressed separately in the tender documentation)

9   Curricula vitae of key persons.
10  Organisational structure.
11  Health and safety plan.
12  Quality assurance plan.
13  Programme.
14  Works Information for *Contractor*-designed *works*.
15  Statement of how the *Contractor* plans to do the work.
16  Design of Equipment.
17  Names of proposed Subcontractors.

This list is made on the basis of a one-stage tender. Where a pre-qualification stage has already been evaluated, there will be no need to request information such as financial statements and statements of previous experience on similar projects, including referees, since these would have been submitted at the pre-qualification stage.

The *Project Manager* should look out for any qualifications in the covering letter or within the tender documents submitted together with any proposed alternatives, etc.

### A1.3.3 Evaluation criteria

It is important to ensure that all the tenders are given due consideration and that a robust tender evaluation process is put in place. Table A1.1 below shows a typical example of an evaluation matrix and it shows some of the criteria upon which an *Employer* may wish to evaluate a tender. (Note again that some of the evaluation criteria listed in the table are

**Table A1.1** Example tender evaluation matrix

| Evaluation criteria | | Weighting (B) | Tenderer No. 1 | | | |
|---|---|---|---|---|---|---|
| | | | Requirements | Score (scores 0 low to 10 high) (A) | Comments | Weighted score $\frac{A}{100} \times B$ |
| | **Technical** | | | | | |
| 1 | Relevant experience | 4.0 | | 8.0 | Good recent experience on similar projects | 0.32 |
| 2 | Technical capability | 20.0 | | 9.0 | Innovative design solutions | 1.80 |
| 3 | Programme | 5.0 | | 7.0 | Some logic incorrect | 0.35 |
| 5 | Subcontractors | 3.0 | Specify what you would expect to see from the tender so that you are able to mark all the tenders in the same way | 6.0 | Quality of Subcontractors generally low | 0.18 |
| 6 | Risk management | 5.0 | | 2.0 | Little consideration given to risk | 0.06 |
| 7 | Quality plan | 5.0 | | 7.0 | Well presented; information incomplete | 0.35 |
| 8 | Health and safety plan | 10.0 | | 7.0 | Well presented | 0.7 |
| 9 | Environmental issues | 3.0 | | 5.0 | Not given enough consideration | 0.15 |
| 10 | Organisational structure and key persons | 2.5 | | 7.0 | Good structure which should marry well with our own team | 0.17 |
| 11 | Partnering philosophy | 2.5 | | 8.0 | Good ethos and outlook | 0.2 |
| | **Commercial** | | | | | 0.0 |
| 12 | *Activity schedule/bill of quantities* | 30.0 | Specify requirements | 9.0 | Robust pricing – no qualifications – all data completed. Competitive pricing | 2.7 |
| 13 | Model compensation event to evaluate percentages | 10.0 | | 7.0 | | 0.70 |
| | Total weighting = | 100.0 | | | | |
| | Total score | | | = 82.0 | | |
| | Total weighted score | | | | | 7.68 |
| | Rank based on weighted score | | | = 3 | | |

selection criteria and some are award criteria. The two sets of criteria should be separated and addressed separately rather than approached together as shown in Table A1.1.

All evaluations are relative to the other tenders; that is, a tender is evaluated as being competitive or not against the backdrop of the market as represented by the tenderers.

With each criterion, you should detail what you are looking for, so that comparisons against the tenderers are made against the same requirements. It is also possible to list requirements as 'essential' and 'desirable' so that those tenderers who do not state that they are capable of providing the items deemed to be essential are dropped from the evaluation.

### A1.3.3.1 Commercial evaluation

The important commercial documents in a tender are as follows.

1 Contract Data part one.
2 *Activity schedule* or *bill of quantities.*
3 Contract Data part two where the *Contractor* inserts key commercial data, for example:
   ■ Full Schedule of Cost Components:
       ■ Working Areas overhead percentage,
       ■ equipment rates,
       ■ manufacture and fabrication overhead percentage,
       ■ design outside of the Working Areas overhead percentage.
   ■ Shorter Schedule of Cost Components:
       ■ percentage for people overheads,
       ■ equipment – percentage for adjustment,
       ■ percentage for design overheads.
   ■ *Direct* and *subcontracted fee percentage.*

Table A1.2 is an example of an evaluation matrix for the entries made in the Contract Data part two by the *Contractor*. For this sort of assessment, the assumptions should be clearly stated, such as the number of hours to be used, or the values on which percentages are to be based. A 'model compensation event' is generally the best way to consider this element of the evaluation. The data for the Schedule of Cost Components can be a large part of the tender evaluation, particularly for Options C, D and E; therefore, careful thought should be given to the assessment of these data.

### A1.3.3.2 Technical

The important technical documents in a tender for the purposes of evaluation include

■ Works Information for *Contractor*-designed *works*
■ programme
■ Statement of how the *Contractor* plans to do the work, resource schedules
■ quality plan.

Table A1.2 is an example of an evaluation matrix for the entries made in Contract Data part two by the *Contractor*.

### A1.4 Conducting the evaluation

The following is a list of items for the *Project Manager* to consider on receipt of a tender. Note again that some items are selection criteria and the two should be addressed separately and not together as indicated in this list.

1 The *activity schedule* is not used to assess change but is more of a cash flow document. Therefore you need to check for front loading of early activities. It should be remembered that early cash flow may reduce the *Contractor*'s finance charges and may enable him to offer a commercially advantageous tender.
2 Key persons – what are the qualifications and experience of the key persons?
3 Programme – how well has the *Contractor* understood the requirements of the client? Does it include all the information requested in the invitation to tender and tendering instructions? Does the programme have any implications for the *Employer*; that is,

**Table A1.2** Example of evaluation matrix for the entries made in Contract Data part two by the *Contractor*

| Tender assessment pro-forma | | | | | | |
|---|---|---|---|---|---|---|
| **Full and Shorter Schedule of Cost Components** | | | | | | |
| 1 Design | | | | | | |
| *Total number of hours to be divided equally between each tendered category = 100 hours* | *Hours* | | *Rate* | *Total* | | |
| Category 1 | ................ | at | | | | |
| Category 2 | ................ | at | | | | |
| Category 3 | ................ | at | | | | |
| Category 4 | ................ | at | | _____ | | |
| Subtotal (B) | = | £ | | _____ | A | |
| 2 Design overheads | ............... % | of (A) | | | = £ ................ | |
| **Full Schedule of Cost Components** | | | | | | |
| 3 Manufacture and fabrication | | | | | | |
| *Total number of hours to be divided equally between each tendered category = 100 hours* | *Hours* | | *Rate* | *Total* | | |
| Category 1 | ................ | at | | | | |
| Category 2 | ................ | at | | | | |
| Category 3 | ................ | at | | | | |
| Category 4 | ................ | at | | _____ | | |
| Subtotal (A) | = | £ | | _____ | B | |
| 4 Manufacture and fabrication overhead | ............... % | of (B) | | | = £ ................ | |
| 5 Working Areas overhead | ............... % | of | | 25 000.00 | = £ ................ | |
| 6 Purchased Equipment with an on-cost charge | | | | | | |
| [insert description] | ................ | X | | [time period] = | | |
| [insert description] | ................ | X | | [time period] = | | |
| **Shorter Schedule of Cost Components** | | | | | | |
| 7 Percentage for people overheads | ............... % | of | | 10 000.00 | = £ ................ | |
| 8 Adjustment for Equipment in published list | ............... % | of | | 10 000.00 | = £ ................ | |
| 9 Rates for the following equipment: | | | | | | |
| [insert description] | ................ | X | | [time] = | | |
| [insert description] | ................ | X | | [time] = | | |
| Subtotal (C) | | | | | = £ ................ | |
| 10 Description of subcontracted work (D) | | | | | | |
| 11 *direct fee percentage* | ............... % | of C − D | | | | |
| 12 *subcontracted fee percentage* | ............... % | of D | | | | |
| **Total of compensation events** | | | | | = £ ................ | |
| **Total for tender assessment** | | | | | = £ ................ | |
| 13 Other factors | | | | | | |
| Completion date (value to *Employer* for every week of time saving) | | | | | | |
| 14 Technical ability | | | | | | |
| Health and safety | | | | | | |
| Quality | | | | | | |
| Design | | | | | | |

does it include dates by which the *Employer* must provide information or 'free issue' materials or do something?

The following is a list of items for the *Contractor* or Subcontractor to consider on issuing a tender.

1 Always ensure that you have fully completed Contract Data part two. Inserting 'non-applicable', 'n/a', 'to be advised', or simply leaving blanks will lead to problems later when you come to discuss compensation events with the *Project Manager*.
2 Remember that the information given in Contract Data part two will be used by the *Employer* to assess your tender and will be used as noted above to assess the effects of change.
3 The percentage for Working Area overheads and the people percentage are used independently of each other.
4 The names of key persons will also be used by the *Employer* to assess your tender.
5 The *activity schedule* (Option A) can be thought of as a milestone payments schedule. You are only paid for these activities when they are complete, so that they may have to be broken down into a number of activities or sub-activities to suit monthly valuations.
6 The Fee comprises a *subcontracted fee percentage* and a *direct fee percentage*, so that the *Contractor* can separate those parts of the *works* done by Subcontractors and those parts done by himself.

**Procuring an Engineering and Construction Contract**
ISBN 978-0-7277-5720-3

ICE Publishing: All rights reserved
doi: 10.1680/pecc.57203.031

# Appendix 2
# ECC tender documentation

**Synopsis**

This appendix provides suggestions for the invitation to tender and ECC tender documentation, including

- Letter of Invitation
- Conditions of Tendering
- tender submission documents
- Contract Data
- Contract Prices
- Works Information
- Site Information.

## A2.1 Introduction

In general, the Invitation to Tender (ITT) and tender documentation should include the following.

- Header page of contract (including name of *Employer*, project/contract number and a description of the *works*).
- Invitation to tender letter.
- Conditions of Tendering.
- Tender submission documents: description of items that the tenderer is required to include with his tender for evaluation purposes.
- Draft form of contract/Articles of agreement.
- Contract Data parts one and two.
- Contract prices.
- Works Information.
- Site Information.

## A2.2 Header page

The header page could include

- name of *Employer*
- description of the *works*
- project title
- project number
- contract number.

## A2.3 Letter of Invitation

The letter could include the following.

- A heading line 'Invitation to Tender for [the *works*]'.
- A statement inviting the tenderer to submit a tender to Provide the Works.
- A statement that the attached documents describe how the tenderer is to submit his tender.
- A statement that a tender submission in response to this invitation to tender is deemed to be acceptance of the tender procedures and *conditions of contract*.
- The name and contact details of the person responsible for issuing the tender documentation.
- The tender return date and time.

## A2.4 Conditions of Tendering

Standard Conditions of Tendering should be included here.

Examples of elements for inclusion are included below; however, these are examples only and should not be read as obligatory inclusions for an ECC contract. If the ECC ITT letter and tender documentation you are issuing is required to take into account the EU Procurement Regulations, some adjustments may need to be made.

### A2.4.1 General

This section could include the following.

- The name and contact details of the person responsible for the invitation to tender.
- The return date and time for the tender.
- A statement that all communication should be in writing.
- A statement that the tender documents at all times remain the property of the *Employer*. The ITT and any associated correspondence are subject to the laws of copyright and must not be reproduced, whether in whole or in part, without the prior written consent of the *Employer*.
- A statement regarding the confidentiality of the ITT and the tender. This is particularly important for the public sector, bearing in mind the Freedom of Information Act 2000.
- A requirement for the tenderer to acknowledge receipt of this enquiry within three working days, clearly stating the intention to accept or decline the invitation. If the invitation is declined, the tenderer should return all the tender documents, which must be treated with total confidentiality, with the acknowledgement.
- A statement that the tenderer shall bear all costs associated with the preparation and submission of his tender.

**A2.4.2 Documentation to be submitted by the tenderer**

This section could include the following.

- A list of the tender documentation included in the ITT.
- A statement that the tender documentation may be modified by the *Employer* prior to the tender return date. Any procedures describing the issuing of addenda should be included as well (e.g. the tender should acknowledge the addendum and the tenderer may ask for an extension of time to the tender return date).
- A statement regarding the acceptability or otherwise of alternative tenders, how this should be effected and any restrictions on the basis of an alternative tender (e.g. different completion date or technical method).

**A2.4.3 Preparation of tenders**

This section could include the following.

- A statement that the tenders should be submitted in the *language of this Contract*.
- A statement that the tenderer should become familiar with all aspects that could affect the price submitted.
- A statement that the prices should be based on the whole of the *works*.
- The validity period of the tender (e.g. 60 days).
- A procedure for dealing with queries (e.g. a time limit on when they can be received, to whom the answers will be given, that all queries and answers should be in writing).

**A2.4.4 Tender submission**

This section could include the following.

- The number of copies of the tender required to be submitted.
- How the tenders should be presented (e.g. in separate envelopes with no markings to indicate the name of the tenderer).
- Whether faxed or emailed copies of the tender will be accepted.
- How late tenders will be dealt with (e.g. returned unopened).

**A2.4.5 Evaluation of tenders**

This section could include the following.

- A statement regarding the *Employer*'s obligations regarding the tender process (e.g. the *Employer* may accept any tender, cancel the tender process, reject all tenders without any obligation to inform the tenderer of the reason for such action, abandon the work or issue another ITT for the same or similar *works* at any time).
- A statement regarding how the *Employer* will award the contract (e.g. the *Employer* does not undertake to accept the lowest tender in respect of the invitation to tender; or the *Employer* shall award the contract on the basis of the most economically advantageous tender in terms of quality of service (weighting %), speed of service (weighting %), technical ability (weighting %) and experience of individuals (weighting %) proposed, together with the method statement (weighting %) as well as anticipated price (weighting %)). (Note that some of the criteria listed are selection criteria rather than award criteria and should be addressed separately by the evaluation team and not all together as indicated in this list.)

**A2.4.6 Evaluation process and post-tender negotiation**

This section could include the following.

- A statement that the *Employer* may enter into contractual negotiations with selected shortlisted tenderers with the intention of entering into a contract with a selected tenderer. The statement could add what this process would entail (e.g. presentation and discussion by the tenderer at a chosen site and/or a meeting at the tenderer's premises to discuss the proposal further and to meet selected personnel).
- A statement that the *Employer* may specify additional requirements or agree detailed conditions of contract as appropriate to the specific requirement with a selected tenderer(s) at any time prior to formal contract award.
- A statement that the contract resulting from the ITT may not be exclusive.

**A2.4.7 Formation of the contract**

This section could include the following.

- The process of notifying tenderers when the evaluation has been completed.

- A statement that unsuccessful tenderers are required, on request, to return all the ITT documents, including any drawings, supplied for the preparation of the tender.
- A statement about how the contract will be formed (e.g. the *Employer* shall issue duplicate original contract documents to the successful tenderer as confirmation of the award of the contract; or the execution of the contract shall be deemed to have taken place when a tender submission and a letter of acceptance have been exchanged between the parties).
- A list of items that are required prior to contract award (e.g. any collateral warranties and guarantees required; any parent company guarantees and bonds required; the policies and certificates for insurance to be provided).
- The process for informing tenderers of the successful contractor and any debriefing offered to unsuccessful tenderers.

## A2.4.8 Contract aspects

This section is optional and could include the following.

- A statement that a particular number of days notice of commencement of site work will be given by the *Project Manager*.

## A2.5 Tender submission documents

This section includes the forms and documents that the *Employer* requires the tenderer to submit with his tender, and could include the following. (Note that some of the items listed are selection criteria not award criteria and so should be addressed separately and not all together as indicated in this list.)

- Form of Tender (including a statement that the *conditions of contract* are accepted).
- Declaration of bona fide tender (if deemed necessary).
- Declaration of site visit (if deemed necessary).
- List of proposed Subcontractors.
- Samples of Materials.
- Working hours for site construction work (these may have already been stipulated in the Works Information).
- Project organisation/management structure, together with curricula vitae of key personnel.
- Health and safety policy and health and safety plan.
- Documentary evidence that the *Contractor* is complying with the Construction (Design and Management) Regulations (the CDM Regulations).
- Industrial relations policy.
- Control of Substances Hazardous to Health Regulations (COSHH) assessments in respect of any hazardous substances used by the *Contractor* for the *works*, whether on or off the site.
- Quality Plan.
- Programme.
- Statement of how the *Contractor* plans to do the work.
- Contractor's Proposal (Works Information for *Contractor*-designed *works*).
- Design of Equipment.
- Haulage routes.
- Design management.
- Environmental policy.
- Risk management.

## A2.6 Contract documents

The next sections are those documents that would be included in any contract resulting from the invitation to tender. The documents would include

- form of contract/Articles of agreement
- Contract Data
- Contract Prices
- Works Information
- Site Information.

**A2.6.1 Form of contract/Articles of agreement**

A form of contract/Articles of agreement document has not been included in this appendix. Each organisation will have their own standard document that should be tailored to suit the ECC terminology. This form could be used to state the Contract Date, which does not appear in other ECC documents.

**A2.6.2 Contract Data**

The Contract Data should be included next. Contract Data part one by the *Employer* should be completed prior to issue, and the choices required to be made in Contract Data part two by the *Contractor* should have been made by the *Employer* prior to issue. See Chapter 3 for a complete listing of the Contract Data entries for the *Employer* and the *Contractor*.

**A2.6.3 Contract Prices**

The *activity schedule* or *bill of quantities* is included here; an example of each is given below. However, these should be preceded by the following.

1   Preamble to the *activity schedule/bill of quantities*.
2   Introduction.

The total of the Prices is fully inclusive for the whole of the *works* and includes all aspects that the *Contractor* is required to do to Provide the Works in accordance with the contract.

Include here instructions for completing the *activity schedule*.

Note that it is not necessary to include lists of things that the Prices should include. The Works Information should make it quite clear what the *Contractor* is to provide, and therefore he should know what he is to price for. In the past, contractors might have said that they had not priced for something because it had not been detailed in the bill or other payment mechanism. In the ECC, this is a spurious argument. Note that the *bill of quantities* or the *activity schedule* is not Works Information.

3a   *Activity schedule*

| Activity number | Description of activity | Lump sum price (£) |
|---|---|---|
| 1 | | |
| 2 | | |
| 3 | | |
| | Total of the Prices | |

or

3b   *Bill of quantities*

| Number | Description | Rate | Unit | Quantity | Lump sum price (£) |
|---|---|---|---|---|---|
| | | | | | |

**A2.6.4 Works Information**

Insert the Works Information here (including the form for parent company guarantee and performance bond). Guidelines for the drafting of the Works Information and what should be included in the Works Information are included in Chapter 4.

**A2.6.5 Site Information**

Insert the Site Information here. Guidelines for the drafting of Site Information and what should be included in the Site Information are included in Chapter 4.

**Procuring an Engineering and Construction Contract**
ISBN 978-0-7277-5720-3

ICE Publishing: All rights reserved
doi: 10.1680/pecc.57203.037

# Chapter 2
# Contract Options

**Synopsis**

This chapter looks at the Contract Options available within the ECC, including

- ECC main and secondary Options
- priced contracts
- target contracts
- cost-reimbursable contracts
- choosing a main Option
- choosing a secondary Option.

## 2.1 Introduction

The main and secondary Options chosen, together with the core clauses, form the terms and conditions applicable to the contract. The choice of main and secondary Options is therefore very important in determining how the *Contractor* is to be paid, where and how the financial risks of the project lie, and what some of the other project risks are.

The Options chosen by the *Employer* form part of the contract strategy for the project. The choice of Options should be reviewed with every contract, rather than simply choosing the same Options that were used in previous contracts. Most contracts tend to have different risks associated with them and different requirements, so that a mix of Options may be more appropriate than opting for the same format every time.

These principles apply equally to contractors and subcontractors when setting up their own contracting strategy.

## 2.2 ECC main and secondary Options – general

There are six main Option clauses (A–F) within the ECC, each of which represents the way in which the *Contractor* will be paid during the period of the contract, as illustrated below in Table 2.1.

**Table 2.1** Main Options (payment mechanisms) under the ECC

| Contract type | Main Option | Price document |
|---|---|---|
| Priced contracts | A<br>B | *Activity Schedule*<br>*Bill of Quantities* |
| Target contracts | C<br>D | *Activity Schedule*<br>*Bill of Quantities* |
| Cost reimbursable | E | Defined Cost |
| Management contract | F | Defined Cost |

In addition to the six main Options, there are 18 different secondary Options, which are subject to some restrictions (Option X1 Price adjustment for inflation may not be used with Options E and F; Option X3 Multiple currencies may only be used with Options A and B; Option X16 Retention may not be used with Option F; Option X20 may not be used with Option X12). In addition, there are two dispute resolution procedure options in ECC3, one of which is required to be chosen: W1 is used unless the UK Housing Grants, Construction and Regeneration (HGCR) Act 1996 applies and W2 is used where the UK HGCR Act 1996 does apply.

| ECC3 Option number | Option |
|---|---|
| X1 | Price adjustment for inflation |
| X2 | Changes in the law |
| X3 | Multiple currencies |
| X4 | Parent company guarantee |
| X5 | Sectional Completion |
| X6 | Bonus for early Completion |
| X7 | Delay damages |
| X12 | Partnering |
| X13 | Performance bond |
| X14 | Advanced payment to the *Contractor* |
| X15 | Limitation of the *Contractor*'s liability for his design to reasonable skill and care |
| X16 | Retention |

| ECC3 Option number | Option |
| --- | --- |
| X17 | Low performance damages |
| X18 | Limitation of liability |
| X20 | Key Performance Indicators |
| Y(UK)2 | The Housing Grants, Construction and Regeneration Act 1996 |
| Y(UK)3 | The Contracts (Rights of Third Parties) Act 1999 |
| Z | Additional conditions of contract |

The headings of Options X8 to X11, X19 and Option Y(UK)1 are not used in ECC3.

The contract document is operational without any of the secondary Option clauses. These are bolt-on provisions, which enable each *Employer* to select provisions in accordance with his individual needs.

Among the secondary Options available are several provisions that in traditional standard forms of contract would be within the main body of clauses, for example, retention and delay damages.

The primary part of the contract strategy is therefore formed by the selection of main and secondary Options and dispute resolution procedure options (Figure 2.1).

**Figure 2.1** ECC contract structure

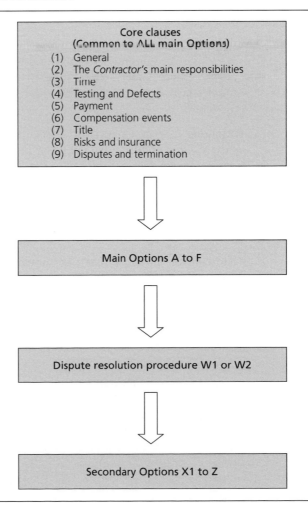

## 2.3 Main Options
### 2.3.1 The choice of main Option

Some of the choices are easy.

- If you want a management contractor, choose Option F.
- If you want a cost-reimbursable contract, choose Option E.
- If you want to use a *bill of quantities*, choose Option B or D.
- If you want to use an *activity schedule*, choose Option A or C, which describes milestone payments.
- If you want a priced contract, choose Option A or B.
- If you want a remeasurement contract, choose Option B.
- If you want a target contract, choose Option C or D.

The selection of a main Option determines the balance of financial risk between the *Employer* and the *Contractor*, as summarised in Figure 2.2. It should also be borne in mind that the objective of a contracting strategy may only be achievable through a combination of the above Options. It may be the case that the combination may require the use of the NEC3 Professional Services Contract, where the *Contractor*/Subcontractor is being asked to be involved at an early stage with design development with the *Employer*.

It should also be noted that the *Contractor*'s and/or Subcontractor's own subcontracts do not necessarily require them to be 'back to back'. For example, the *Contractor* or Subcontractor may be on an Option C contract. He can himself have subcontracts let on Option A.

Some employers in partnering/framework arrangements require the *Contractor*'s or Subcontractor's own contracts to be 'back to back'; for example, if the *Contractor*'s contract is Option C then so should the Subcontractor's be. This often occurs where the employer wishes to see a flow down through the supply chain of incentivisation on a project. Stating your partnering approach/philosophy and how you intend to approach incentivisation of your own supply chain are often questions asked as part of the tendering process.

> A *Contractor* appointment on an Option C Target Cost with *activity schedule* contract does not necessarily have to have an Option C subcontract. Unless, of course, it is a requirement of the *Employer*'s contract.

The choice of main Option also affects the *Contractor*'s incentive and the flexibility enjoyed by the *Employer* during the contract. Fixed-price contracts tend to increase the risk to the *Contractor* and decrease the risk to the *Employer*. Fixed-price contracts also require a more fixed scope, however, and consequently the *Employer* has less flexibility available to him. In addition, if the fixed scope has to change because it was not properly defined, the *Employer*'s risk increases. The graph in Figure 2.3 shows the flexibility, risk and incentive patterns that characterise the types of contracts available within the ECC.

Following the principle that risk should be placed with those best able to manage it, the choice of main Option could depend on the quality of the Works and Site Information available at tender stage. Put simply, a *Contractor* faced with poor-quality information

**Figure 2.2** Balance of risk for each main Option under the ECC

| Contract type | Main Options | Balance of risk | |
|---|---|---|---|
| | | *Employer* | *Contractor* |
| Priced | A / B | | |
| Target | C / D | | |
| Cost reimbursable | E | | |
| Management contract | F | | |

Figure 2.3 Characteristics of main Options in the ECC

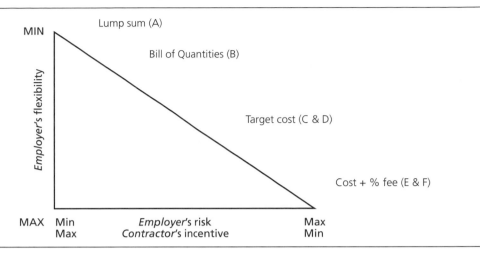

while tendering for a priced contract may include a significant allowance for risk. In these circumstances the *Employer* will subsequently pay the *Contractor* the tendered sum whether or not the risk arises. Where the scope of work is poorly defined, the *Employer* is well advised to select a cost-reimbursable option, as by doing so he will pay only if the risk occurs.

Other factors influencing the choice of main Options include where an *Employer* might feel that a target contract could result in a more open relationship with the *Contractor* and that successive contracts using Option C could result in a price reduction through the use of incentives (using Option C even when the scope is clear may facilitate a more open and beneficial relationship).

Many employers tend to prefer priced contracts (Options A and B – note that Option B is not a fixed-price contract in the same way as Option A, since Option B is fully remeasurable), where the budget for the contract is relatively secure, barring variations ordered by the *Employer* or his representative. In order to receive a fixed-price contract from a *Contractor*, however, the scope of works is required to be well defined and fixed. This is often not possible. In this case, the *Employer* has a decision to make about whether he wants to take the risk of changes made to the contract, or whether he wants to share the risk with the *Contractor*, who could make valuable contributions to the solving of buildability and design issues. This could change his choice of main Option from A to C or from B to D.

In some cases, an *Employer* may have an idea about what a target contract (Options C and D under the ECC) represents. Some employers may regard a target contract as a means of 'sharing the risk'. This is, of course, an aspect of any target contract, but there are other considerations as well. Principally, the behaviour of the two parties to the contract should be modified from the arms-length attitude of a fixed-price contract, to the more collaborative relationship that is required to make a target contract work.

A primary driver for the choice of main Option under the ECC is the quality and standard of the Works Information available. The Works Information describes the *Employer*'s requirements. An Option A contract based on a Works Information that is only 50% complete may give rise to numerous changes that will change the profile of the contract from a fixed-price low-*Employer*-risk contract to a variable-price and therefore higher-risk contract.

**2.3.2 Trade-offs in choosing a main Option**

The choice of main Option can also depend on the trade-off between cost, time and quality (see Figure 2.4). For example, a contract for which the point in the triangle is close to the time apex could mean that the time of completion (e.g. Millennium Dome or shops/retail outlets) or the starting time (e.g. annual factory shutdowns or limited working windows) is

**Figure 2.4** Project management trade-off of cost, time and quality

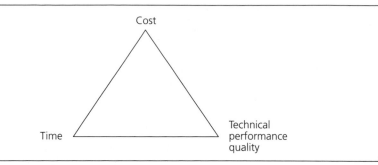

more important than the price. This does not mean that price is **not** important, rather that time is more important, and therefore price-bound Options such as A and B might not be the appropriate choice in this situation.

It is vital to choose the right main Option in the first instance. Problems could arise where a main Option is chosen that does not match the completeness of the Works Information or other parts of the contract strategy. Where there are so many compensation events occurring on an Option A contract that the *Project Manager* is struggling to keep abreast of them all, this may be an indication that the Works Information is incomplete and that the contract should have been an Option C/D or E contract.

According to the ECC guidance notes (notes in brackets are the authors'), the following are factors that should be taken into account when deciding which main Option to choose.

- Who has the necessary design expertise? (This could affect the main Option under the assumption that Options A and B are designed by the *Employer*; however, a design-and-build contract could be completed using Options A or B with a few adjustments.)
- Is there a particular pressure to complete quickly? (Pressure to complete quickly would usually lead to an Option E contract.)
- How important is performance of the completed *works*?
- Is certainty of final cost more important than lowest final cost (pointing towards an Option A contract rather than Option C)?
- Where can a risk best be managed? (The *Contractor* carries more financial risk in Options A and B and less risk in Option E.)
- What total risk is tolerable for contractors?
- How important is cross-contract coordination to achievement of project objectives? (In this case, Options C, D or E would be more suitable.)
- Does the *Employer* have good reasons for selecting specialist contractors or suppliers for parts of the *works*?

The clarity of the *Employer*'s objectives is important in choosing the correct main Option for the project. It is not possible to achieve all of certainty of price, certainty of completion date, the ability to change the works, reallocating risk to the *Contractor*, retaining design responsibility and lowest price. It is important to ensure that the specification is correct where there is *Employer* design.

> Risk should be placed with those best able to manage it.

## 2.4 Priced contracts

There are two priced Options available in the ECC.

1. A: Priced contract with *activity schedule* (where the *Contractor* is paid on completion of activities).
2. B: Priced contract with *bill of quantities* (a remeasurable contract).

### 2.4.1 A: Priced contract with *activity schedule*

An *activity schedule* (called an Activity Schedule with concomitant definition) is a list of activities prepared by the *Contractor* that represents the activities that he expects to undertake in carrying out the *works*. In traditional contracts, these activities might have been called milestone payments.

#### 2.4.1.1 A1: Tender

Option A requires the tenderer to tender lump sum prices against the *activity schedule*. The Prices are the lump sums attached to each activity and the total of the Prices is the contract sum.

If the *Contractor* is given a totally free hand to produce the *activity schedule*, this could make it difficult for the *Employer* to evaluate tenders where each tenderer submits an *activity schedule* containing different activities (where the invitation to tender letter and accompanying tender documentation were part of a competitive tender exercise). If the *Employer* is overly prescriptive in the detailing of the *activity schedule*, this could affect the *Contractor*'s cash flow, where the *Employer* has not detailed a sufficient number of activities.

In an attempt to find a balance between the two, some *Employers* include guidelines or state requirements on the minimum level of breakdown for the *activity schedule* or activities within it in much the same way as a contract sum analysis or work breakdown structure. This enables the *Employer* to identify significant items of work or stages of work that he is particularly interested in or concerned about and enables these items to be clearly identified, and also facilitates greater ease of comparison between tenderers during tender evaluation. Such requirements should be detailed in the instructions to the tenderers at the time of tender.

For example, the *Employer* could draw up a list of primary activities that will form the basis of his evaluation exercise, and instruct the tenderers to price these activities and any other sub-activities that they wish to add that make up the primary activities. For example, the *Employer* could list the following as primary activities in the *activity schedule* where the *works* is an office building: access routes, foundations, floor, walls, roof and fit-out.

A counter-argument to all the above and the reason why the ECC guidance notes suggest that the *Contractor* prepares the *activity schedule* is because a *Contractor* may be able to offer a commercially advantageous tender through the structuring of payments due for completed activities.

The *activity schedule* in its purest form is an opportunity for the *Contractor* to structure his payments and cash flow to suit his requirements and be able to offer the benefit of this through his tender to the *Employer*.

Concerns are often raised in relation to front loading of early activities by the *Contractor*. Some employers, especially those with budgetary constraints on yearly spend, are interested in and may seek options/alternative prices from tenderers where they are able to offer them early or up-front payments for work subject to certain conditions outlined in additional *conditions of contract* in an Option Z clause.

#### 2.4.1.2 A2: Risk

Because Option A is a lump sum contract, there tends to be more risk allocated to the *Contractor* than to the *Employer*. The *Contractor* is paid for his work based on the lump sum prices tendered against the *activity schedule*. He therefore carries the risk for accurate pricing for the contract overall (in just the same way as he would in any other lump sum contract) and his cash flow is based on the allocation of the lump sum price against each activity in the *activity schedule*.

#### 2.4.1.3 A3: Payment

Under Option A, the *Contractor* is paid for activities that are complete at each assessment date. The amount to be paid is the amount entered by the tenderer against that activity in the *activity schedule* (as amended for compensation events).

There are two points to note here. The first is that the activity is required to be complete by the assessment date if it is to be included in the amount due. This means that the *Contractor*

**Figure 2.5** Activities spanning assessment dates

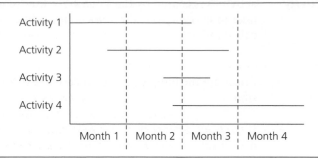

could have programmed and included the activity so that it would meet an assessment date. Any change in the programme could therefore affect this programme and the *Contractor*'s cash flow. The second is that an activity is complete when it has reached completion in accordance with the contract. In general, this definition will be included in the Works Information and should be objective enough for the *Project Manager* to decide when the activity is complete. (For a fuller description of Completion, see Chapter 1 of Book 1 and Chapter 2 of Book 3.)

It is therefore important that the duration of activities described is not so long that they span assessment dates, since the completed activities are assessed at each assessment date. In Figure 2.5, activity 1 will be assessed as complete at the end of month 3 and will therefore be paid at the end of month 4 (or thereabouts). The *Contractor* will be required to finance this cash flow from the beginning of the project until the end of month 4 – a heavy burden indeed. Activities 2 and 3 will also be assessed as complete at the end of month 3 and activity 4 will be assessed when it is complete.

A balance needs to be struck between

- the contractual requirement for each activity to be represented on the programme and
- the need to ensure that there should not be so many activities that the programme becomes cumbersome and too onerous to be revised/reviewed every month.

In Figure 2.6 below, there are many activities which may prove a challenge to manage. Cash flow is assured.

The most effective solution is to programme activities so that they can be completed before an assessment date. There may be situations where it is necessary to amend the definition of the Price for Work Done to Date in clause A11.2(27) to allow for percentages of activities completed, so that the correct balance of the number of activities in the programme and *activity schedule* is achieved (see Figure 2.7).

**2.4.1.4 A4:**
**Compensation events**

In theory, few changes should take place in an Option A contract since it supposedly has a well-defined scheme of works and it is a lump sum contract. Small compensation events

**Figure 2.6** Activities too small to be included in the programme

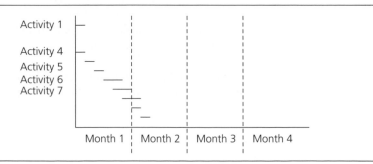

**Figure 2.7** Activities designed to suit the assessment dates

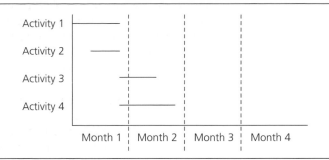

could still occur, however. In this case, the assessment for any compensation events that are implemented will be added to the lump sum price of the relevant activity, thereby increasing the total of the Prices. The payment taking place on completion of that activity will therefore be the tendered lump sum adjusted for compensation events.

Compensation events are priced using the Shorter Schedule of Cost Components. Rate or lump sum may also be used to assess a compensation event instead of Defined Cost (clause A63.14).

A compensation event may delay completion of an activity so that completion occurs one or more assessment dates later than the *Contractor* planned at tender stage. This could affect the *Contractor*'s cash flow since payment for that activity could be a month or more later than expected.

Consideration should also be given by the *Employer* and *Contractor* to any work being identified as a **new activity** which would de-link the completion of this additional/extra work from the completion of the whole activity.

Compensation events are priced using only the Shorter Schedule of Cost Components rather than the *activity schedule*. It is important therefore that the *Contractor* completes the data for the Schedule of Cost Components in Contract Data part two. In the past, some tenderers have left this section blank, thinking that if it is a lump sum contract, there is no requirement to provide data for the Schedule of Cost Components (the theoretical ramifications of this are that the *Contractor* carries out compensation events free since no data have been provided or inserted by the *Contractor*). In this instance, there could be room for negotiation (although the *Project Manager* could refuse because the Parties have already entered into contract); and certainly disagreement between the Parties. It is more effective and expedient for the *Employer* and *Contractor* to ensure that these data have been completed in the first place.

**2.4.1.5 A5: Selection**

A fuller discussion of the aspects to be taken into account when choosing a main Option is included in section 2.2 above. In general, Option A should be selected by the *Employer* when

- the Works Information is complete (does not necessarily imply correct)
- a firm price and low risk are more important than flexibility.

**2.4.2 B: Priced contract with *bill of quantities***

A *bill of quantities* is a list of work items and quantities with the unit of measurement as stated in a *method of measurement*, and which describes the items to be included and how the quantities are to be calculated.

**2.4.2.1 B1: Tender**

Option B requires the tenderer to tender rates and prices against each item included in the *bill of quantities*. The *bill of quantities* is usually prepared by the *Employer*. The sum of all products (rate × quantity) is the total of the Prices.

**2.4.2.2 B2: Risk**

Option B is a priced contract; therefore, there tends to be more risk allocated to the *Contractor* than to the *Employer*. The *Contractor* is paid for his work at the prices tendered against the *bill of quantities* and therefore he carries the risk of having priced the items correctly, although he does not carry the risk of changes in quantity.

Some *Employers* have amended Option B contracts so that the *Contractor* is to price for the risk associated with changes in quantities. This amendment effectively makes it a lump sum priced contract like Option A, but using a *bill of quantities*.

**2.4.2.3 B3: Payment**

Option B is a remeasurable contract. The *Contractor* is paid for the actual quantity of work completed, as identified in the *bill of quantities* multiplied by the rates and prices against each of these items.

Because the *Contractor* is paid according to what he has completed in the period up to the assessment date, there is no requirement for completion of activities as there is with Option A.

**2.4.2.4 B4: Compensation events**

Compensation events are priced using the Shorter Schedule of Cost Components rather than the *bill of quantities*, although if both parties agree, a rate or lump sum may be used by agreement to assess a part of a compensation event instead of Defined Cost (clause B63.13). The default position (of pricing compensation events using the Shorter Schedule of Cost Components) is to overcome the problems associated with rerating or star rates for *bill of quantities* items. It is aimed at preventing the *Employer* being disadvantaged where the quantities in the *bill of quantities* were incorrect and some loading of the *bill* has taken place. It also protects the *Contractor* from the *Project Manager* insisting on inappropriate rates and prices in the *bill of quantities* being used to assess the cost implications of the change.

As with Option A, it is important therefore that the *Contractor* completes the data for the Schedule of Cost Components in Contract Data part two.

Note that there are three extra compensation events (clauses B60.4, B60.5 and B60.6) in Option B, which recognise that it is a remeasurable contract and that the final remeasurement at the end of the contract might reveal inconsistencies that apply retrospectively.

**2.4.2.5 B5: Selection**

A fuller discussion of the aspects to be taken into account when choosing a main Option is included in section 2.2 above. In general, Option B should be selected by the *Employer* when

- a firm price is required but some changes in quantities may occur, for example, road maintenance
- the *Employer* or his professional team wishes to use a *bill of quantities*
- a schedule of defined works or services is required, for example, term maintenance contract.

## 2.5 Target contracts
### 2.5.1 Adjustment of the target price (Options C and D only)

A target price contract works as follows. At tender stage the *Contractor* assesses the cost of doing the defined work in the same way he would do under any other contract arrangement. Having added his fee for overhead and profit and made any other tender adjustment in the usual way, the price he arrives at and submits with his tender constitutes the target price or the 'total of the Prices' as the ECC refers to it (not to be confused with the definition Total of the Prices in Option D, which is used as a comparison with the Price for Work Done to Date (PWDD) for the purpose of calculating the *Contractor's* share). Depending on whether Option C or D is used, the total of the Prices is either expressed by reference to an *activity schedule* (Option C) or a *bill of quantities* (Option D).

During the course of construction the *Contractor* is paid his own Defined Cost, which he has already paid out plus the Fee (referred to collectively as the 'Price for Work Done to Date'). When the *works* are complete the relationship between the final PWDD and the target price determines whether the '*Contractor's share*' is positive or negative (see Figure 2.8).

**Figure 2.8** *Contractor's* share

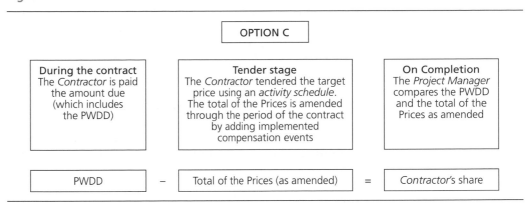

In order to make this arrangement equitable to the parties and maintain the incentive to 'out-perform' the target price, it will be necessary to adjust the target price when compensation events occur. Their effects on cost (and time) are assessed in exactly the same way under the ECC cost-based options as they are under the price-based contracts (Options A and B).

### 2.5.2 Disallowed Cost

The big concern for *Employers* with any cost-based contract is the risk that, in the absence of the moderating and controlling effect of agreed rates and prices for the work associated with price-based contracts, the *Contractor* will be less inclined to control costs and consequently the *Employer* will not obtain good value. It is for this reason that most cost-based contracts, the ECC included, incorporate some 'protection' for the *Employer* from 'excessive' inefficiency on the part of the *Contractor*. This protection works by identifying circumstances in which certain costs will be excluded from Defined Cost so that the *Contractor* does not get paid for them. The ECC achieves this by grouping together a list of *Contractor* shortcomings under the definition of Disallowed Cost (clause 11.2(25) for Options C, D, E and clause 11.2(26) for F).

When calculating the *direct fee percentage* to be included with his tender submission, the *Contractor* will tend to make an assessment of the likely risk of Disallowed Cost arising and reflect this in the *direct fee percentage*.

### 2.5.3 The *Contractor's* share (Options C and D only)

The principal difference between a target contract and a cost-reimbursable contract is that the former attempts to address the possibility of excessive inefficiency on the part of the *Contractor* by introducing the incentive of a further payment to the *Contractor* if he can complete the *works* for a cost **less** than the target price (as adjusted in accordance with the terms of the contract).

Conversely, if the final cost of the *works* exceeds the target price it is usual with target contracts for a sum of money to be deducted from the amount due to the *Contractor*. It follows that any target contract needs to be very clear at the outset about how this further payment (to or from the *Contractor*) is to be assessed and when it becomes due.

The *Employer* should be aware that some contractors, aware that their inefficiencies could result in a smaller share, could inflate the target price in the first place, by tendering inflated Prices. In a competitive tender situation, this behaviour should be minimised; however, it is still a possibility.

Although simple in concept, at first sight the provisions in the ECC take some understanding, particularly clause 53.1 (clause C53.1; D53.5), which deals with the calculation of the 'further payment' referred to as the *Contractor's share* by the ECC. The way it works is as follows.

### 2.5.3.1 Deciding on the share

The *Employer*, when preparing Contract Data part one, decides how he wishes to 'share' with the *Contractor* any under- or over-run of the final cost against the target price. The extent of the financial risk to the parties in the event of the final cost **exceeding** the target price can be varied between two extremes.

1  A guaranteed maximum price to the *Employer* can be achieved by stating the *Contractor's share percentage* to be 100%. (In simple terms the *Contractor* gets paid no more than the target price and is left to absorb totally the excess over the target (referred to as the guaranteed maximum price.)

2  A minimum fee to the *Contractor* can be achieved by stating the *Contractor's share percentage* to be 0%. (In simple terms the *Contractor* will always get paid his Defined Cost plus an additional Fee.) In essence this becomes a pure cost-reimbursable contract.

In order to understand how the *Contractor's share* is calculated under the ECC target options, an example is given below which shows a range of possible outcomes based on particular '*share ranges*' and '*Contractor's share percentages*' selected by the *Employer*.

When deciding how to apportion the financial risk between *Employer* and *Contractor* by means of the share percentages, the two key factors are

1  the degree of confidence in the target price submitted and

2  the incentive must be of sufficient value to produce the desired effort and economy (i.e. it should be designed to encourage savings in Defined Cost rather than encourage a temptation to exaggerate increases in the target price).

Since the idea is to provide an incentive to the *Contractor*, it may be wise not to have a high *Contractor's share percentage* for low *share ranges* or for high *share ranges*. Where, for example, the *Contractor's share percentage* is 75% for a *share range* of less than 80%, the *Contractor* could be incentivised to tender high Prices, rather than incentivised to be more efficient. It is perhaps more practical to suggest a range so that the *Contractor* is incentivised for a Price for Work Done to Date (PWDD) just below the total of the Prices, but less incentivised for a PWDD far below the total of the Prices.

An example of *share ranges* and *Contractor's share percentage* that aims to maximise the *Contractor*'s incentive is given in Table 2.2.

In this example, if the PWDD is over 20% less than the target price (total of the Prices), then the *Contractor* is paid 20% of this under-run. He is paid less of the under-run in this instance than if the PWDD were between 10% and 20% less than the total of the Prices. This is based on the suggestion that if the difference is greater than 20%, the total of the Prices was incorrect and the smaller share incentivises the *Contractor* to be more accurate about his tendered total of the Prices. Conversely, if the PWDD is more than 20% greater than the total of the Prices, the share of that over-run required to be paid by the *Contractor* is only 10%. This will hopefully still leave the *Contractor* with a little profit. Many *Employers* do not hold with this latter philosophy and instead insert either 100% (as a guaranteed maximum price), or something greater than 50% as a disincentive to the *Contractor* reaching beyond the total of the Prices.

Table 2.2  Example of *share ranges* and *Contractor's share percentage*

| Share range | Contractor's share percentage |
| --- | --- |
| Less than 80% | 20% |
| From 80% to 90% | 40% |
| From 90% to 110% | 50% |
| From 110% to 120% | 25% |
| Greater than 120% | 10% |

**2.5.3.2 Paying the share**

The payment of the *Contractor's share* is made in two stages.

1  At Completion of the whole of the *works* based on a forecast by the *Project Manager* at that time of the likely final out-turn cost and the likely final target price.
2  Later when the actual final out-turn cost and the actual target price are agreed.

If between the above two dates the *Project Manager*'s earlier forecast proves to be incorrect, the *Project Manager* has an obligation to assess the new amount due and certify a further payment.

The *share* is only calculated at Completion even though the *Contractor* could have been paid more than his *share* by the time Completion occurs (which would mean the *Contractor* having to pay back his share to the *Employer*). The ECC2 guidance notes give two reasons why interim payments of the *Contractor*'s share are not provided for before Completion of the whole of the *works*.

1  The Prices tendered by the *Contractor* (either in the form of an *activity schedule* – Option C, or a *bill of quantities* – Option D) have the main purpose of establishing the original target price. It is not intended that their build-up should provide a realistic forecast of cash flow and they are unlikely to be comparable with the PWDD (Defined Cost plus Fee) at any interim stage.
2  Forecasts of both the final PWDD and the final total of the Prices would be extremely uncertain at early stages of the contract. Any delays in assessing compensation events would further distort the calculation.

The danger of serious under- or over-payment of an interim *Contractor*'s share has therefore led to the policy of an estimated payment on Completion, which is corrected when the assessment of the final amount due is made.

**2.5.3.3 Calculation of Contractor's share**

Note that the following examples are for the purposes of this section only and do not form recommended ranges.

Information entered by the *Employer* in Contract Data part one:

| Share range | Contractor's share percentage |
|---|---|
| Less than 90% | 50% |
| From 90% to 120% | 25% |
| Greater than 120% | 10% |

If at Completion of the whole of the *works* the *Project Manager* forecasts that the final total of the Prices (called the Total of the Prices (as defined) in Option D) (i.e. the adjusted target price) will be £10 million then the above table becomes:

| Final PWDD[a] (*share range*) | Contractor's share percentage |
|---|---|
| Less than £9m[b] | 50% |
| From £9m to £12m[c] | 25% |
| Greater than £12m | 10% |

[a]The PWDD is defined in clause C and D 11.2(29)
[b]Where £9m is 90% of the forecast final total of the Prices of £10m
[c]Where £12m is 120% of the forecast final total of the Prices of £10m

Given the above information (the **total** of the Prices – the Total of the Prices (as defined) in Option D – also known as the target price, remains the same for all these examples, at £10 million), below are examples of possible outcomes. Remember that more than one *share range* may be used per example. (This is because the calculation is always taken from the middle point up or down. Where the PWDD < total of the Prices (see examples (a) and (b)), the calculation is made from the range within which the total of the Prices lands and continues down through all the other ranges below that. Where the PWDD > total of the Prices (see example (c)), the calculation is made from the range within which the total of the Prices lands and continues up through all the other ranges above that.)

---

**Example (a)**                                                     **Total/total of the Prices = £10m**

**Final Price for Work Done to Date = £8m**

In this example, the amount paid to the *Contractor* during the period of the contract was £8m. The total of the Prices as tendered by the *Contractor* and adjusted for compensation events was £10m. This £10m, also known as the target price, forms the base comparator for all calculations.

The final PWDD at £8m is less than the adjusted total of the Prices at £10m. In this instance, the calculation will breach two *share ranges* as follows. The *Contractor's share percentage* of 50% (less than £9m – see above) will be applied for the amount from £8m to £9m, the parameter for that *share range*. The *Contractor's share percentage* of 25% will be applied for the amount from £9m to £10m, which falls within the parameter of the second *share range*.

| | |
|---|---|
| Saving under final total of the Prices | = £2m comprising two *share ranges* (where £2m = £10m total of the Prices – £8m PWDD) |
| Less than £9m | = £1m × 50% = £0.50m (£1m is the difference between £9m, which is the top of that *share range*, and £8m, the PWDD) |
| From £9m to £12m | = £1m (£1m is the difference between £9m, which is the top of the previous *share range* and £10m, which is the total of the Prices) × 25% = £0.25m |
| *Contractor's* share paid by *Employer* | = £0.750m |
| Total amount due to the *Contractor* (excl. VAT) | = £8.75m |
| | This £8.75m is calculated by adding the £8m already paid to the *Contractor* to the £0.75m now due to him as the *Contractor's* share. |

---

**Example (b)**                                                     **Total/total of the Prices = £10m**

**Final Price for Work Done to Date = £9.5m**

In this example, the amount paid to the *Contractor* during the period of the contract was £9.50m. The total of the Prices as tendered by the *Contractor* and adjusted for compensation events was £10m. This £10m, also known as the target price, forms the base comparator for all calculations.

The final PWDD at £9.5m is less than the adjusted total of the Prices at £10m. In this instance, the calculation will use one *share range* only. The *Contractor's share percentage* of 25% will be applied for the amount from £9.50m to £10m, which falls within the parameter of the second *share range*.

| | |
|---|---|
| Saving under final total of the Prices | = £0.50m comprising one *share range* (where £0.5m = £10m total of the Prices – £9.5m PWDD) |
| From £9m to £12m | = £0.50m × 25% = £0.125m (£0.5m is the difference between £10m total of the Prices within the same *share range* as the £9.5m PWDD) |

Contractor's share paid by Employer = £0.125m
Total amount due to the Contractor = £9.625m
(excl. VAT)

> This is calculated by adding the £9.50m already paid to the Contractor to the £0.125m now due to him as the Contractor's share.

---

**Example (c)**                                    **Total/total of the Prices = £10m**

**Final Price for Work Done to Date = £11m**

In this example, the amount paid to the Contractor during the period of the contract was £11m. The total of the Prices as tendered by the Contractor and adjusted for compensation events was £10m. This £10m, also known as the target price, forms the base comparator for all calculations.

The final PWDD at £11m is more than the adjusted total of the Prices at £10m and there has been an over-run of the target price. In this instance, the calculation will use one share range only. The Contractor's share percentage of 25% will be applied for the amount from £10m to £11m, which falls within the parameter of the second share range.

Excess over the final total of the Prices = £1m comprising one share range (where £1m = £11m PWDD − £10m total of the Prices)
From £9m to £12m = £1m × 25% = £0.25m (£1m is the difference between the £11m PWDD within the same share range as the £10m total of the Prices)
Contractor's share paid to Employer = £0.25m
Total amount due to the Contractor = £10.75m
(excl. VAT)

> This is calculated by subtracting the amount of £0.25m now due by the Contractor from the £11m already paid to the Contractor.

---

This method of calculating the Contractor's share is used for both Option C and Option D contracts.

## 2.5.4 C: Target contract with *activity schedule*

An *activity schedule* (defined as the Activity Schedule in Option C) is a list of activities that is expected to be carried out by the Contractor in carrying out the works.

### 2.5.4.1 C1: Tender

Option C requires the tenderer to tender lump sum prices against an *activity schedule* that may have been constructed by the Employer or by the Contractor, depending on the instructions given by the Employer at tender stage (see A1 in section 2.4.1.1 above). This represents the target price (total of the Prices) at which the Contractor estimates he can do the work described in the contract.

### 2.5.4.2 C2: Risk

The financial risk of the project is shared by the Contractor and the Employer, and target contracts therefore offer a middle-of-the-road option in terms of risk allocation. All other general and project risks are as allocated in the contract.

Option C is a popular choice of option by employers who wish to share risk. It allows a risk-and-reward approach that incentivises both parties to work together. The cost of the Employer's changes are visible and the Contractor is incentivised to be efficient and to assist the *Employer* in completing the project on time and to budget. It could also be that some *Employers* find it difficult to provide a complete Works Information before the work starts and Option C allows them to start the work sooner and for the project to continue smoothly.

### 2.5.4.3 C3: Payment

The *Contractor* is not paid in accordance with the *activity schedule*.

During the period of the *works*, the *Contractor* is paid his Defined Cost plus his Fee (clause C11.2(29)). In other words, the *Contractor* is paid the amounts due to Subcontractors (without taking into account various amounts deducted for default) plus the amount due in accordance with the Schedule of Cost Components less Disallowed Cost plus any profit and overhead (see Chapter 2 of Book 4 on the Schedule of Cost Components for a fuller explanation). He is also paid his Defined Cost plus his Fee for each compensation event. The tendered Activity Schedule is not used for payment.

### 2.5.4.4 C4: Compensation events

The Prices tendered by the *Contractor* in the *activity schedule* are used as a target price for the project (clause C11.2(30)); that is, the total of the Prices is used as the base comparator for the amount paid during the period of the contract (represented by the PWDD). By the nature of the payment option chosen, it is likely that the *works* will change and that compensation events will take place. The changes to the Prices quoted by the *Contractor* for implemented compensation events are used to change the Prices in the *activity schedule* (clause C63.12) so that the target price continually keeps pace with the changes to the *works*. In this way the *activity schedule* (the target price) will keep in line with the actual job and will form a realistic target at Completion.

### 2.5.4.5 C5: Share

At Completion (clause C53.3), the total of the Prices – that is, the amount represented in the *activity schedule* – is compared with the Price for Work Done to Date – that is, the total amount paid to the *Contractor*. The difference between this target price (total of the Prices) and the amount paid to the *Contractor* (PWDD) is shared between the *Employer* and the *Contractor* in a predetermined way (as stated in Contract Data part one).

If the amount paid to the *Contractor* is greater than the target price, then the *Contractor* pays his share of the difference. If the amount paid to the *Contractor* is less than the target price, then the *Contractor* receives his share of the difference (see section 2.5.3.3 above for an example).

### 2.5.4.6 C6: Other considerations

There are potential difficulties with this method of payment, however.

*Employers* may envisage a difficulty in retrieving at Completion money that has already been paid to the *Contractor* as PWDD, where the PWDD exceeds the total of the Prices, especially if retention was not chosen as a secondary Option, or if retention does not cover the amount due.

Although retention is an obvious answer to a concern about retrieving an over-run of target cost from the *Contractor*, it does not reflect the real reason for using retention. Another way of alleviating this concern is to amend clauses to change the time when the difference between the Price for Work Done to Date and the total of the Prices is first assessed. This can be done, for example, as soon as the PWDD exceeds the total of the Prices, or two months before Completion is expected.

However, there are a number of difficulties with this approach.

- The administration of the contract has to be absolutely accurate and up to date in order to ensure that the total of the Prices fully reflects all compensation events, otherwise the basis of the comparison would be incorrect. Depending on the number of changes taking place, this might be quite difficult, particularly taking into account the passage of time in the compensation event procedure.
- The calculations taking place every assessment date could become quite onerous, since both the PWDD and the total of the Prices will continue to change as time goes

by. It is only when the work is complete that any difference can be assessed with any fairness, since anything can happen in contracts, and frequently does.

In general, therefore, it is recommended to rely on the contract and on the *Contractor* and to assess the risks of getting it wrong before changing the contract. Note that the definition of PWDD in Option C and D clause 11.2(29) includes for a forecast of the Contractor's payments before the next assessment date.

**2.5.4.7 C7: Selection**

A fuller discussion of the aspects to be taken into account when choosing a main Option is included in section 2.3 above. In general, Option C should be selected when

- the scope of the work cannot be fully defined or detailed, or is uncertain
- a reliable budget can be produced and the effects of risk can be best managed by a combined *Employer* and *Contractor* team working together to meet the *Employer*'s objectives.

Option C can be used to form the contractual basis of a partnering approach.

**2.5.5 D: Target contract with *bill of quantities***

Option D follows the same principle as Option C except that a *bill of quantities* is used to reach the tendered target price instead of an *activity schedule*. Usually the *bill of quantities* is drafted by the *Employer*.

A fuller discussion of the aspects to be taken into account when choosing a main Option is included in section 2.3 above. In general, Option D should be selected by the *Employer* when

- the extent of the work is not fully defined
- a reliable budget can be produced and the effects of risk can be best managed by a combined *Employer* and *Contractor* team working together to meet the *Employer*'s objectives
- the *Employer* or his professional team wishes to use a *bill of quantities*.

**2.6 Cost-reimbursable contracts**

**2.6.1 E: Cost-reimbursable contract**

Option E is at the other end of risk allocation to Options A and B. In Options A and B, the *Employer*'s flexibility is low compared with the *Contractor*'s incentive to complete on time and to budget. In Option E, the *Contractor* is paid all properly expended costs and therefore his incentive is lower, although the *Employer*'s flexibility is high.

The *Contractor* is paid his Defined Cost plus his Fee during the period of the contract (clause E11.2(29)). There is no target price to compare the costs against and therefore compensation events are of less importance from a cost perspective (some *Employers* manage an Option E contract in the same way as Option C or D, constantly comparing the PWDD against an internal budget, adjusted for compensation events. This might assist the *Employer* in managing his budget, but this activity is extremely onerous given that one of the reasons for choosing Option E is that the scope is not fully developed and many compensation events are expected), although they still impact the programme. Where the *Employer* has his own budget that he wishes to manage, some cost estimate may be required from the *Contractor*, such as the contractual requirement to provide forecasts of Defined Cost on a regular basis (clause E20.4), and compensation events may therefore acquire a renewed importance.

A fuller discussion of the aspects to be taken into account when choosing a main Option is included in section 2.3 above. In general, Option E should be selected by the *Employer* when

- the definition of the work to be done is inadequate even for a target contract and
- an early start to construction is nevertheless required, for example, safety critical, emergency repair works.

**2.6.2 F: Management contract**

A management contractor contracts directly with Subcontractors and manages them on behalf of the *Employer* (see Figure 2.9).

**Figure 2.9** Management contract

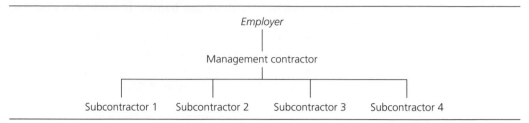

ECC3 recognises that the management contractor may also undertake some works himself. The *Contractor* is required to indicate in Contract Data part two the work which he proposes to undertake himself.

Note that Option F is not for use as a construction management contract, where the employer contracts with all contractors including the construction manager who manages the works, but does no part of the works himself. In this case, it is suggested that the employer contract with the construction manager using the NEC3 Professional Services Contract. An alternative is to purchase a licence from the publishers of the contract and amend the *conditions of contract* to reflect projects where more than one contractor is working on the site. The terms and conditions would need to reflect the greater communication and coordination required and could also include elements such as set-off between the many contractors on the site.

A fuller discussion of the aspects to be taken into account when choosing a main Option is included above. In general, Option F should be selected by the *Employer* when

- the *Employer* sees time/cost/performance benefits from using the experience and services of a management contractor to assist in the management of a high-risk project
- the *Employer* does not have the internal resources or experience to manage a project.

**2.7 Administering the ECC cost-based contract**

When using a cost-based contract (target price or cost-reimbursable contracts such as Options C, D, E and F), *Employers* tend to have two main doubts about their ability to control their financial commitments. The first arises from the difficulty of predicting the final cost; the second is a lack of incentive for the *Contractor* to control cost and to complete the work expeditiously.

Under a target/cost-reimbursable contract, financial control therefore takes on far greater emphasis given the increased financial risk to the *Employer* under such contractual arrangements. Consequently, those main Option clauses specific to the ECC Options C, D, E and F that **do not** address how amounts due (payments) to the *Contractor* are assessed, concentrate on ensuring that the *Employer* (through the *Project Manager*) is afforded the opportunity for greater involvement in any matter that could impact upon the *Contractor*'s Defined Cost and therefore how much the *Employer* is liable to pay.

It is possible to address the lack of incentive to control cost and to complete the work expeditiously by introducing bonuses or Key Performance Indicators which reward the *Contractor* for such items as

1 quality and robustness of project reporting through:
- accurately forecasting the final out-turn cost of the project
- audit of Defined cost
- programme/time management
2 bonuses for early completion.

**2.7.1 Greater financial control by the *Employer***

This section addresses how the ECC provides for the *Employer* to exercise greater control over matters which could impact upon Defined Cost.

**2.7.1.1 Clause 20.3**      Clause 20.3 (for main Options C, D, E and F) states that the *Contractor* advises the *Project Manager* on the practical implications of the *Employer*'s design of the *works* and on subcontracting arrangements. Although such obligations are typical for cost-based contracts and reflective of the assumed closer degree of cooperation and openness between the parties, there is little guidance given as to how detailed the advice should be and no contractual sanctions applied to the *Contractor* who fails to enter into the spirit of the provision.

Value management including engineering provisions could be added into the ECC, for example, to allow the *Contractor* to identify a design he believes to be more economic and practical to construct. The effect of such a provision, if more economic designs proposed by the *Contractor* are adopted, is that both the *Employer* and the *Contractor* benefit financially in proportion to any share mechanism included. This is a provision designed to encourage the *Contractor* to maximise his share potential (profit) while at the same time reducing the overall cost to the *Employer*.

**2.7.1.2 Clause 20.4**      Clause 20.4 (for main Options C, D, E and F) emphasises the need for the *Employer*, through the *Project Manager*, to be more financially aware under cost-based contractual arrangements. It states that:

> 'The *Contractor* prepares forecasts of the total Defined Cost for the whole of the *works* in consultation with the *Project Manager* and submits them to the *Project Manager*. Forecasts are prepared at the intervals stated in the Contract Data from the *starting date* until Completion of the whole of the *works*. An explanation of the changes made since the previous forecast is submitted with each forecast.'

Although there is no clear contractual purpose for the forecasts, if prepared carefully they are vitally important to financial awareness and cost control, since they will identify any deviations from the *Contractor*'s financial assumptions included with the original tender. To maximise the benefit of these forecasts it is recommended that the following steps are taken:

1   Request tenderers in their tender submission to give a full breakdown of their tender price, showing all their assumptions regarding resource levels, outputs, input costs, subcontracting arrangements and risk allowances. This information is essential for two purposes. First, it allows the *Employer*, as part of the tender assessment, to determine whether the tender submitted is realistic. Second, and more relevant to cost forecasting, it effectively provides the first forecast of the total Defined Cost for the whole of the *works* and therefore is the yardstick for all subsequent forecasts.
2   Include in the Works Information a description of the minimum level of detail and the format required for the forecasts. This is important because the ECC requirement for forecasts is very broad in terms of the detail required and, if no further guidance were to be provided in the Works Information, forecasts which serve no useful purpose may result. All of these forecasts essentially have two components as follows.
    ■ The costs committed in providing the *works* that have been completed up to the date that the forecast is prepared.
    ■ The current estimate to complete the remaining *works*.

When preparing the Works Information, it should be ensured that these forecasts yield reliable information that can be used to identify problem areas requiring investigations, but at the same time utilising as far as possible any existing cost monitoring/control systems the *Contractor* has in place. More sophisticated techniques such as Earned Value Analysis can be used to assist in the preparation of the forecasts, but the golden rule must be to keep the forecasts as simple as possible consistent with providing reliable information. Any forecast of the total Defined Cost of the whole of the *works* will only be as reliable as the accuracy of the following elements of it.

■ The measurement at the forecast date of the physical work done.
■ The accuracy of the overall costs committed in completing the work done and, just as important, the overall **allocation** of those costs to the appropriate activities.

- The accuracy of the estimate of costs for the work remaining to be done at the forecast date.

If all this sounds like hard work, it is worth remembering that the success or failure of a contract is closely related to the managerial effort expended by the parties administering the contract.

Construction planning is largely concerned with the efficient use of labour and constructional plant. In any type of cost-based contract the *Employer* pays for most or all of these resources on an actual cost basis. Since the cost of such resources is mainly time-related the *Employer* should concern himself with this aspect of contract management. There is consequently a strong case on cost-based contracts for a joint cost and planning team comprising staff from the *Contractor*'s and the *Project Manager*'s team working from the same offices producing revised programmes (as required by clause 32.2), forecasts of the total Defined Cost for the whole of the works (clause 20.4) and quotations in respect of any compensation events that arise (as required by clause 62.3).

Forecasts are prepared at intervals stated in the Contract Data. Many contractors only prepare such forecasts internally on a three-monthly basis. The greater financial risk accrued by the *Employer* on a cost-based contract justifies such forecasts being prepared on a more frequent basis.

## 2.7.1.3 Subcontracts

It was stated above that Defined Cost as defined includes payments the *Contractor* makes to Subcontractors in accordance with the terms of the latter's subcontract. It follows therefore that the *Project Manager* should have a greater awareness of the commercial details of subcontracts and may wish to include in the Works Information procurement procedures to be followed by the *Contractor* for all subcontracted work (over, say, a stipulated value) and likewise for supply contracts. The safeguards provided by the ECC in respect of work the *Contractor* elects to subcontract are as follows.

1 Before appointing a proposed Subcontractor the *Contractor* must submit the name to the *Project Manager* for acceptance. The *Contractor* is not permitted to appoint a proposed Subcontractor until the *Project Manager* has accepted him (clause 26.2) and doing so is grounds for termination (clause 91.2 (R13)).

2 Before appointing a proposed Subcontractor the *Contractor* must submit the proposed conditions of subcontract to the *Project Manager* for acceptance (clause 26.3), unless either a contract in the NEC suite of documents is to be used or the *Project Manager* has agreed that no submission is required. Again the *Contractor* is not permitted to appoint a proposed Subcontractor on the proposed subcontract conditions until the *Project Manager* has accepted them (clause 26.3).
This is an important provision in a cost-based contract, allowing the *Project Manager* to at least have the opportunity to comment on the proposed financial and contractual risk allocation between the *Contractor* and the Subcontractor. For example, the *Employer* may not be amenable to the payment of a large 'premium' under a lump sum subcontract arrangement for, say, the transfer of the unforeseen physical conditions risk to the Subcontractor.
Furthermore, given that the definition of Disallowed Cost includes costs which the *Project Manager* decides 'should not have been paid to a Subcontractor in accordance with his subcontract', it is clearly important that the *Project Manager* has knowledge of the detailed terms and conditions of all subcontracts.

3 Specific to the cost-based options of the ECC, clause 26.4 requires the *Contractor* to submit the proposed Subcontract Data for each subcontractor to the *Project Manager* for acceptance if instructed to and (rather obviously) if a contract in the NEC contract series is to be used. Strangely enough the submission of this is not mandatory, but of more concern is **not** a condition precedent to the appointment of the Subcontractor.
Given the all too common occurrence of a failure by contractors to understand the Schedule of Cost Components and its related commercial percentages (entered by the Subcontractor in part two of the Subcontract Data), it is the Subcontract Data that

probably represents the greatest likelihood of the *Employer* paying 'over the odds' on a cost-based contract.

In summary, subcontracted work represents a specific risk to the *Employer* under an ECC cost-based contract and it is recommended that careful consideration is given to the inclusion in the Works Information of procurement procedures that ensure the *Project Manager* is kept advised throughout.

## 2.7.2 Assessing and verifying the *Contractor's* Defined Cost

This section addresses how, under the ECC, amounts due to the *Contractor* are assessed and, given the cost-based nature of these sums, how the *Employer* verifies that such costs have been properly incurred by the *Contractor*.

In any cost-based contract, both parties are primarily concerned with Defined Cost – that is, that cost as defined (clauses C11.2(23), D11.2(23), E11.2(23), F11.2(24)) and incurred by the *Contractor* in carrying out the work. Consequently, it is essential that the *Project Manager* starts to think in terms of cost and the manner in which construction costs are incurred rather than the unit price approach more commonplace with the majority of price-based contracts. The *Contractor's* accounts and records for the contract must be open to scrutiny by the *Project Manager*, who will need to establish monitoring and auditing procedures to ensure that these documents and records constitute an accurate reflection of the costs properly incurred by the *Contractor* in providing the *works* defined in the contract. Audit should proceed and be completed concurrently with the construction work.

There is no express mention of 'audits' by the *Project Manager* in the ECC. Instead, and sensibly, the ECC merely describes in general terms the accounts and records that the *Contractor* is obliged to keep (clause 52.2) and which he must give the *Project Manager* the facility to inspect (clause 52.3). It is for the *Project Manager* to determine the nature and frequency of his 'inspections' of the *Contractor's* accounts and records, with the objective of providing confidence to the *Employer* that the latter is only being required to pay those amounts to which the *Contractor* is entitled in accordance with the terms and conditions of the contract.

Consistent with providing the necessary 'confidence', audits should be designed to achieve maximum 'value' from minimum effort. They should be structured and focused, recognising that in some cases 80% of the *Contractor's* total Defined Cost will be represented by 20% of the cost components listed in the Schedule of Cost Components. They should concentrate on the 'higher-risk' cost components such as subcontracted work, internal plant hire and 'muck-away'. They should be carried out concurrent with the construction work so that their findings are able to support as closely as practicable the assessment by the *Project Manager* of amounts due to the *Contractor*. An example audit plan is included in Appendix 3 below. Exhaustive audits are not always necessary; random audits may suffice.

> On a major infrastructure project, a detailed audit plan was prepared by the project team for the tunnels and stations contract, which left the *Contractor* in no doubt as to the nature and frequency of the audits to be carried out by the *Project Manager* and the information, in the form of accounts and records, that he was expected to provide. Despite some initial resistance, once the audits were under way the *Contractor* admitted that some of the audit findings were beneficial in identifying where it was necessary to 'tighten up' his internal procedures.

Below is an example of the detailed checks that should form part of any audit of the *Contractor's* costs under an ECC cost-based contract. To practitioners more familiar with the unit pricing more common with price-based contracts, the auditing of a *Contractor's* costs can seem rather daunting. It is therefore useful to remember that on any construction contract all of the *Contractor's* costs fall into one of the following categories.

■ Direct costs of production.

- Indirect costs of production (site overheads/on-costs).
- Risk allowances not included above.
- Head office overheads.
- Profit.

In turn, each of these (with the exception of the last two) constitutes the cost of one or more of the following components.

- People (labour and staff).
- Construction plant.
- Materials.
- Subcontractors.
- Miscellaneous other charges (e.g. rents payable for temporary occupation of land).

It should not be surprising then that the Schedule of Cost Components, which is the list of admissible costs constituting Defined Cost for **non-subcontracted works**, is broken down into similar cost headings, namely

- people
- Equipment (construction plant and other temporary works)
- Plant and Materials (things incorporated into the permanent works)
- charges
- manufacture and fabrication outside the Working Areas
- design done outside the Working Areas
- insurances.

The audit when carried out is seeking to ensure the following.

- Only Defined Cost as defined by reference to the Schedule of Cost Components is included in amounts due to the *Contractor*.
- Cost components not listed in the Schedule of Cost Components are not included in the calculation of Defined Cost. Given that all of the *Contractor*'s costs, which are not listed in the Schedule of Cost Components, are deemed to be covered by the Fee (clause 52.1), if such costs are also taken into Defined Cost there will effectively be a 'double' payment.
- The cost of components notionally covered by a percentage mark-up on a well-defined cost component (e.g. certain overhead costs covered by the percentage for Working Areas overheads) is also not directly reimbursed and again therefore leads to a 'double' payment.
- Any Disallowed Cost as defined (main Options C, D, E clause 11.2(25), and Option F clause 11.2(26)) is identified.

**2.7.3 Payment procedures**

The principal component of amounts due (payments) to the *Contractor* under the ECC is the Price for Work Done to Date, defined (clause 11.2(29) of Options C, D, E and F) as 'the total Defined Cost which the *Project Manager* forecasts will have been paid by the *Contractor* before the next assessment date plus the Fee'.

Expanding this definition, the amount due to the *Contractor* comprises

- amounts paid by the *Contractor* to Subcontractors for work which is subcontracted, less any Disallowed Cost
- amounts paid by the *Contractor* in respect of any of the components listed in the Schedule of Cost Components for work which is not subcontracted, less any Disallowed Cost, and
- the Fee calculated by applying the *subcontracted fee percentage* to the first bullet above and the *direct fee percentage* to the second bullet above.

The important point to draw from the above definition is that interim amounts certified to the *Contractor* only include amounts **paid (not incurred)** by the *Contractor* up to the date that the *Project Manager* assesses the amount due. Consequently, costs committed by the *Contractor* and invoices received but not paid at the assessment date are excluded.

The result of this is that a *Contractor* on an ECC cost-based contract, unless amended, will always be in a negative cash flow position. There are three potential ways of managing this negative cash flow.

1  Since it is not always easy to identify where a *Contractor* has actually paid for things, some clients have introduced into their contracts an obligation on the *Contractor* to set up at the Contract Date a dedicated contract bank account in the joint names of the *Employer* and the *Contractor*, through which all payments to and made by the *Contractor* flow. Such an arrangement has the added benefits of:
   - streamlining the assessment by the *Project Manager* of interim amounts due to the *Contractor* and
   - providing the 'starting-point' for the audit by the *Project Manager* of the *Contractor*'s Defined Cost.

   To obtain the maximum benefit from the operation of such a joint bank account, consideration should be given to amending the Schedule of Cost Components to ensure that Defined Cost as defined effectively mirrors the **exact** costs incurred by the *Contractor*. These amendments essentially involve the elimination of the various percentage 'mark-ups' as included in the Schedule of Cost Components (e.g. the percentage for Working Areas, overheads, and the percentage for Equipment depreciation and maintenance). Such amendments, if carefully thought out, can significantly reduce the administration time spent separating from the *Contractor*'s accounts and records those cost components directly reimbursed and those notionally covered by a percentage mark-up.
2  There is no provision in ECC3 to recover finance charges.

The PWDD includes a forecast of the Defined Cost the *Contractor* will have paid by the next assessment date. This therefore should ensure that the Contractor's cashflow remains cash neutral.

Under the ECC the *Project Manager* is required to **certify** payments to the *Contractor* within one week of each assessment date, failure to do so resulting in the *Employer* becoming liable to pay interest on the late certified amount. If for no other reason, this provides the motivation for the *Project Manager* to ensure he has access to the necessary information to allow him to speedily assess the amounts due. For his part, the ECC (in clause 52.2) obliges the *Contractor* to keep

- accounts of his payments of Defined Cost
- proof that payments have been made
- communications about assessments of compensation events for Subcontractors
- other records as stated in the Works Information.

Clause 52.3 further requires the *Contractor* to allow the *Project Manager* to inspect at any time within working hours the accounts and records which he is required to keep.

## 2.8 Dispute resolution procedure Options in ECC3

In ECC3, dispute resolution procedure Options are contained in Options W1 and W2. Option W1 is to be used where the Housing Grants, Construction and Regeneration Act 1996 (HGCR) does not apply; that is, in countries other than the UK, and in the UK where the contract is not a construction contract under the Act. Option W2 is to be used in the UK where the HGCR applies to the contract. Note that the Local Democracy, Economic Development and Construction Act 2009 amends some parts of the HGCR Act and the NEC has been amended accordingly.

Chapter 4 of Book 3 discusses dispute resolution in detail. A brief description of Options W1 and W2 is included below.

### 2.8.1 Option W1

Option W1 is to be used when the HGCR does not apply. The dispute resolution procedure describes adjudication as the first level of dispute resolution, with the *tribunal* (usually litigation or arbitration) as the second level of dispute resolution. Adjudication is not statutory in this case; therefore, an alternative dispute resolution procedure could easily be incorporated into the contract through the use of Option Z. An example is including adjudication as a later level of dispute resolution, with the first levels being internal escalation procedures.

The primary supposition made within Option W1 (and, indeed, Option W2) is that the Parties will appoint the *Adjudicator* under the NEC3 Adjudicator's Contract at the *starting date* (clause W1.2(1)). It is unlikely that most Parties will undertake this task; they are more likely to include in the Contract Data an adjudicator nominating body to name an *Adjudicator* for appointment when the need arises (i.e. when the Parties fall into dispute and adjudication is required). Option W1 does in fact allow for an *Adjudicator nominating body* (identified in the Contract Data) to choose an adjudicator if the originally chosen one is not identified in the Contract Data, or if he resigns or is unable to act (clause W1.2(3)).

There is also an adjudication table included in W1.3 which sets out the matter which may be referred to the *Adjudicator* by the *Employer* and/or the *Contractor*.

The rest of Option W1 describes the procedures to be followed when a dispute has been referred, and includes the procedures for referring a dispute to the *tribunal*.

### 2.8.2 Option W2

Dispute resolution procedure Option W2 is to be used only in the UK when the HGCR applies to the contract. The Option describes the procedure for statutory adjudication, as dictated by the HGCR. A second level of dispute resolution, the *tribunal* (usually litigation or arbitration), is also included in this Option.

The assumption made within Option W1 with regard to appointing the *Adjudicator* under the NEC3 Adjudicator's Contract at the *starting date* (clause W2.2(1)) is also made in Option W2. It is unlikely that most Parties will undertake this task; they are more likely to include in the Contract Data an adjudicator nominating body to name an *Adjudicator* for appointment when the need arises (i.e. when the Parties fall into dispute and adjudication is required). Option W2 does in fact allow for an *Adjudicator nominating body* (identified in the Contract Data) to choose an adjudicator if the originally chosen one is not identified in the Contract Data, or if he resigns or is unable to act (clause W2.2(3)).

The most important aspect of Option W2 to highlight is that a Party may refer a dispute to the *Adjudicator* at any time (clause W2.1(1)). This is required by the Act, and it is onerous on the *Employer*.

The rest of the Option details the procedures required to be followed in the event of an adjudication under Option W2.

## 2.9 Secondary Options
### 2.9.1 The selection of secondary Option

The main Option chosen by the *Employer*, together with the chosen secondary Options and the dispute resolution procedure options, describe the contract strategy for the particular project required. With the few exceptions that are outlined in the schedule of options on page 1 of the ECC (option X3 may not be chosen with Options C, D, E and F; Option X1 may not be used with Options E and F; Option X16 may not be used with Option F; X20 may not be used with X12), the *Employer* may choose any, all or none of the secondary Options.

One of the principles behind this pick-and-mix approach is that some of the secondary Options may increase the cost of the project. In some traditional contracts, for example, retention and a performance bond are written into the standard conditions of contract, whether the *Employer* requires these aspects of the contract or not. The *Employer* therefore pays for these aspects even if he does not want them.

The flexible approach of the ECC towards contract elements such as liquidated damages, retention and performance bonds means that the *Employer* has a choice about what aspects of the contract are important to him, and what he is prepared to pay for.

**2.9.1.1 Option X1 – price adjustment for inflation (only used with Options A, B, C and D)**

The addition of a price fluctuation clause to main Options E and F is unnecessary, as it is a fully cost-reimbursable contract. Indeed under main Options A–D, all change is valued at Defined Cost, therefore automatically allowing for any and all price fluctuations. However, Option X1 requires the adjustment of compensation event assessments to return to base date levels. Subsequently, the Price Adjustment Factor is applied to the whole amount due at each assessment date.

When Option X1 is included, adjustments are made in accordance with the additional defined terms for which values are required in Contract Data part one as follows.

- The Base Date Index (B) is the latest available index before the *base date*.
- The Latest Index (L) is the latest available index before the date of assessment of an amount due.
- The Price Adjustment Factor is the total of the products of each of the proportions stated in the Contract Data multiplied by $(L - B)/B$ for the index linked to it.

The total of the Prices for the contract are divided into a series of proportions, for example

- substructure
- superstructure
- M&E services
- external works.

Each is linked to an index published by a source also identified and containing the price fluctuations for the appropriate element of work. Provision is made in the Contract Data to identify a proportion of the *works,* which remains non-adjustable.

The clause goes on to define the administration of the Price Adjustment Factor separately for Options A and B and Options C and D.

**2.9.1.2 Option X2 – changes in the law**

The inclusion of Option X2 makes a change in the *law of this contract* a compensation event if it occurs after the Contract Date. The *Employer* carries the risk of all resultant cost and time effects.

In the absence of this secondary Option, under general law the time effects of changes in the law are at the *Contractor*'s risk. This Option would generally be used for longer-term contracts where changes in the law may be unpredictable.

**2.9.1.3 Option X3 – multiple currencies (used only with Options A and B)**

Two identified terms are fundamental to this Option for which the *Employer* in part one of the Contract Data enters values as

- *currency of this contract*
- *exchange rates.*

Contract Data part one will also contain a list of the items and activities which will be paid in a currency other than the *currency of this contract,* along with the currency and the maximum amount to be paid in that currency. Amounts above the maximum amount for a listed activity will be paid in the *currency of this contract.*

Also relevant to Option X3 are the contents of clause 51.1, which states that all payments are in the *currency of this contract* unless otherwise stated in this contract. For the cost-reimbursable main Options C and D clause 50.6, Options E and F clause 50.7 states that payments of Defined Cost made by the *Contractor* in another currency are paid to him by the *Employer* in the same currency. For calculation of the Fee and any *Contractor*'s share, the payments are converted to the *currency of this contract.*

**2.9.1.4 Option X5 – sectional completion**

Sectional completion provides for the *Employer* to take over *sections* of the *works* as they are completed.

This Option operates where invoked by redefining all references in the contract conditions to the *works*, Completion and the Completion Date. These terms subsequently apply to either the whole of the *works* or any *section* of the *works*. Note that the *defects date* still runs from Completion of the whole of the *works*.

Through Option X5 separate Completion Dates are set for the *sections* of the *works* stated in Contract Data part one. Additionally, Contract Data part one allows for sectional Completion Dates to be related to bonuses under Option X6 and delay damages under Option X7.

### 2.9.1.5 Option X6 – bonus for early completion

This Option puts in place a bonus to be paid to the *Contractor* for completing the *works* earlier than the Completion Date (with Options C, D and E, the *Contractor* could expend more (and the *Employer* pay more) in order to earn the bonus).

The amount per day is stated in Contract Data part one and is multiplied by the number of days from earlier Completion (or the date on which the *Employer* takes over the *works* whichever is the earlier) until the Completion Date shown on the latest Accepted Programme.

When Options X6 and X5 (sectional completion) are invoked together, an amount per day for each section of the *works* is stated.

> A contract to build a large casino in an entertainment complex included a bonus for early Completion of £40 000 per day. The owner of the casino knew that he could make at least £40 000 per day every day that the casino was open.

### 2.9.1.6 Option X7 – delay damages

This option contains the converse provision to Option X6 in that the *Contractor* pays delay damages at the amount per day for each day that Completion (or the date on which the *Employer* takes over the *works*) is later than the Completion Date as shown on the latest Accepted Programme.

The delay damages invoked through Option X7 are liquidated damages, the purpose of which is twofold

- to provide compensation to the *Employer*
- to limit the liability of the *Contractor* for failing to complete the *works* by the required date.

Through legal precedent, liquidated damages must be a genuine pre-estimate of the *Employer*'s loss or a lesser sum. The inclusion of a greater amount in Contract Data part one as a penalty will render the provisions of Option X7 unenforceable under law. The *Employer* must be able to demonstrate, if required, that the calculations were done before the start of the contract and that they estimated the *Employer*'s losses in the event that the *Contractor* completed later than required.

In the absence of Option X7, the *Employer* is entitled by law to pursue the *Contractor* for damages at large. It is important to remember that this right is removed by including Option X7 and the *Employer* becomes entitled only to the amount per day stated in the Contract Data. Careful consideration must be given to the value entered.

Again as with Option X6, Option X7 can be invoked in conjunction with Option X5 (sectional completion) and separate amounts per day for delay damages can be stated in the Contract Data for each *section* of the *works*.

Clause X7.3 describes the procedure to reduce delay damages in the situation where the *Employer* takes over a part of the *works* before Completion. The delay damages are reduced immediately from the date of take over of the part of the *works* and the delay damages are reduced by the same proportion assessed by the *Project Manager* as the benefit to the *Employer* of taking over that part of the *works* rather than all the *works* not previously taken over.

**2.9.1.7 Option X12 – Partnering Option**

Option X12 is broken into a number of key components as follows.

- Identified and defined terms clauses X12(1) to (5). This section defines the Partners, Own Contract, the Core Group, Partnering Information and Key Performance Indicators.
- In X12.2 Actions – spells out what actions the Partners will take.
- X12.3 Working together – how the Partners will work together.
- X12.4 Incentives – sets out how much a Partner is paid and what additional payments he may receive for meeting any key Performance Indicators.

In neither Option is there a definition/identification of the *Client*'s representative, who leads the Core Group unless otherwise stated in the Partnering Information. Presumably it is the *Project Manager*; however, it could be any member of the *Client* who can be construed as representing the *Client*.

**2.9.1.8 Option X13 – performance bond; and Option X4 – parent company guarantee**

Like many other options, bonds cost the *Contractor* money that he is likely to pass on to the *Employer*. The reasoning for choosing such options should be carefully considered (see Chapter 3 below on completing the Contract Data for more information). Option X13 and Option X4 should not be chosen together but as alternatives.

These options require the *Contractor* to complete bonds and parent company guarantees in the form stated in the Works Information. This provides the flexibility for the individual project to specify a form of bond or guarantee appropriate to their individual needs. The amount for a performance bond is stated in the Contract Data part one by the *Employer*.

If the bond is considered important then the following additions could be made to the contract.

- It could be stated in the conditions of tendering that the bond is required before execution of the contract.
- An Option Z clause could be added to ensure that the bond is received, for example: 'One quarter of the Price for Work Done to Date is retained in assessments of the amount due until the *Contractor* has submitted the performance bond required under Option X13 to the *Project Manager* for acceptance.'

**2.9.1.9 Option X14 – advanced payment**

Where the *Contractor* is required to make a large capital investment such as marble tiling or the procurement of plant and machinery or special equipment and needs financial support in order to procure the materials at the beginning of the contract, the *Employer* could choose to lend the *Contractor* this money as an advanced payment. Another way of achieving this type of payment is through the *activity schedule* (through adding an activity in the *activity schedule* (Option A) for the purchase of the material so the *Contractor* is paid on completion of that activity) or the *bill of quantities* (through adding an item in the *bill of quantities* that can be assessed at the assessment date).

The information required in the Contract Data for operation of this Option is as follows.

- Amount of advanced payment.
- Period after which repayment instalments begin.
- Amount of instalments.
- Requirement for advanced payment bond.

Where such an advanced payment is made, the amount is stated in Contract Data part one and the *Employer* makes the payment within four weeks of the Contract Date or (if an advanced payment bond is required) within four weeks of the later of the Contract Date and the date when the *Employer* receives the advanced payment bond.

The *Employer* may choose whether he wants an advanced payment bond or not. A bank or insurer which has been accepted by the *Project Manager* issues an advanced payment bond. The *Project Manager* may reject the proposed bondsman if its commercial position is not strong enough. The Works Information stipulates the form of the bond and Contract Data part one stipulates the amount of the instalments which the *Contractor* pays the *Employer* in each amount due following the period stated in Contract Data part one. The bond is for the amount of the advanced payment which the *Contractor* has not repaid.

### 2.9.1.10 Option X15 – design liability limitation

In the absence of this secondary Option, the *Contractor*'s liability for Defects in the *works* due to his design is 'fitness for purpose'. In simple terms this means that the finished *works* must fulfil the function for which they were intended.

The inclusion of Option X15 in the contract conditions limits the standard of liability to reasonable skill and care, bringing it in line with the liability placed on professional designers. This is particularly useful in design-and-build contracts where the contractor may be unable to purchase fitness-for-purpose insurance for his design.

It is important to remember several key points.

- The limitation applies to Defects arising as a result of the *Contractor*'s design and not to the design itself.
- The *Contractor*'s compliance with 'fitness for purpose' or 'reasonable skill and care' can only be determined by reference to the *Employer*'s requirements as stated in the Works Information.
- Through section 4 of the core clauses, the *Contractor* is required to correct Defects before the end of the *defect correction period*, which commences at Completion for Defects notified before Completion and when the Defect is notified for other Defects. The liabilities placed on the *Contractor* by secondary Option X15 cannot therefore be acted upon by the *Employer* until after Completion of the *works*.

X15.2 makes it clear that the *Contractor*'s correction of a Defect for which he is not liable under this contract is a compensation event.

### 2.9.1.11 Option X16 – retention

Retention is a fundamental issue that has traditionally formed a part of the main body of clauses in most conventional contracts. The reasons for requesting retention should be carefully considered since it costs the *Contractor* money to forego a part of the money due to him every month. This reminds the *Employer* of the need for a careful review of the secondary Options to ensure that the desired provisions form part of the contract document.

Retention with Option X16 allocates a *retention-free amount*, where for some period at the beginning of the project the *Employer* will not withhold an amount from the *Contractor*, smoothing the *Contractor*'s cash flow when he needs it most. If the *Employer* wishes to use retention in the traditional manner, he may enter an amount of nil as the *retention-free amount*. It follows that if the *retention-free amount* is an amount equal to 50% of the total of the Prices at the Contract Date, then retention is taken at double the usual amount to accumulate to the same amount at Completion.

An example is given in Figure 2.10.

Two items of information must be included in the Contract Data to operate Option X16.

1 The *retention-free amount* (intended to assist the *Contractor's* cash flow during the early stages of the contract).
2 The *retention percentage* (applied to all amounts due in excess of the above).

On amounts due, the *Employer* retains the *retention percentage* applied to the Price for Work Done to Date exceeding the *retention-free amount* until the earlier of the Completion of the whole of the *works* or the date on which the *Employer* takes over the whole of the

Figure 2.10 Retention

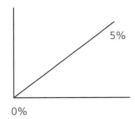
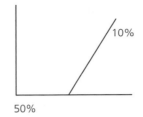
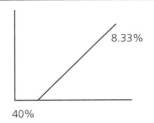

A *retention-free amount* of 0% of the total of the Prices at the Contract Date means that retention will be deducted from the first assessment of the amount due and that the *retention percentage* can be a typical 5%.

A *retention-free amount* of 50% of the total of the Prices at the Contract Date means that retention will start to be deducted once the value of the *works* are approximately 50% complete. This means that the *retention percentage* should be increased to 10% to arrive at the same retention amount at Completion.

A *retention-free amount* of 40% of the total of the Prices at the Contract Date means that retention will start to be deducted once the value of the *works* are approximately 40% complete. This means that the *retention percentage* should be increased to 8.33% to arrive at the same retention amount at Completion.

*works*. This amount is halved at that point and the remainder is paid to the *Contractor* following issue of the Defects Certificate.

**2.9.1.12 Option X17 – low-performance damages**

This Option states that where a Defect included in the Defects Certificate shows low performance with respect to a performance level stated in the Contract Data, the *Contractor* pays the amount of low-performance damages stated in the Contract Data.

In order to ensure that this option operates effectively where invoked, the *Employer* must ensure that the Defect is measurable against information in both the Works Information and the Contract Data. These types of requirements are usual with process plant, for example, water treatment, power stations, factories.

> If, for example, the Defect is in relation to the electrical consumption of a heating system, the performance specification in the Works Information must be clear and provide all necessary values such as the heating levels to be achieved against the acceptable levels of electricity use. The statements given in the Contract Data regarding the amount of damages and the corresponding performance level must be linked to the Works Information in a way that allows the identification of 'low performance'.

**2.9.1.13 Option X18 – limitation of liability**

There are five Contract Data entries required in Contract Data part one. Because the inclusion of the clauses and associated Contract Data part one statements restrict liability to the degree stated, it may be read into the contract that their exclusion results in liability being unlimited.

Clause X18.1 restricts the *Contractor*'s liability to the *Employer* for the *Employer*'s indirect or consequential loss. The amount of the restriction is stated in the Contract Data: 'The *Contractor*'s liability to the *Employer* for the *Employer*'s indirect or consequential loss is limited to …'.

It should be noted that there is no limit for direct losses unless a cap on the total liability is stated in clause X18.4.

Clause X18.2 restricts the *Contractor*'s liability to the *Employer* for loss of or damage to the *Employer*'s property. The amount of the restriction is stated in the Contract Data: 'For any one event, the *Contractor*'s liability to the *Employer* for loss of or damage to the *Employer*'s property is limited to …'.

One should note the term 'for any one event'; the sum stated here is therefore not a total liability but a total liability for any one event. These words are consistent with the wording usually found in insurance policies for *Employer's* property.

Clause X18.3 restricts the *Contractor's* liability to the *Employer* for Defects due to his design that are not listed on the Defects Certificate. The amount of the restriction is stated in the Contract Data: 'The *Contractor's* liability for Defects due to his design which are not listed on the Defects Certificate is limited to . . .'. There is no reason why the Contract Data sentence could not read 'The *Contractor's* liability for Defects due to his design that are not listed on the Defects Certificate is unlimited.' This clause covers what are effectively latent Defects.

Clause X18.4 restricts the *Contractor's* total liability to the *Employer* except for the excluded matters listed in the clause. The excluded matters are amounts payable by the *Contractor*, such as

- loss or damage to the *Employer's* property
- delay damages if Option X7 applies
- low-performance damages if Option X17 applies and
- *Contractor's* share if Option C or Option D applies.

Some *Employers* include in the list of excluded matters such additional items as

- death or personal injury to any person,
- wilful default, fraud or abandonment by the *Contractor*.

The amount of the restriction is stated in the Contract Data: 'The *Contractor's* total liability to the *Employer* for all matters arising under or in connection with this contract, other than the excluded matters, is limited to . . .'. If an *Employer* does not include this clause in the Contract Data, there is a possibility that a *Contractor* may request that it is included. It is possible that an interpretation of its absence by choice could be that the liability is unlimited.

Clause X18.5 restricts the *Contractor's* liability to the *Employer* to a specific period of time after the Completion of the whole of the *works*. The statement in Contract Data part one reads: 'The *end of liability date* is . . . years after Completion of the whole of the *works*.' Once again, *Contractors* may require this clause to be included in their contracts.

An *Employer* may not wish to incorporate all five limitations into his contract. In such a situation, it is suggested that only the relevant clauses are included in the Contract Data. For example, in the list of Options in the first entry in Contract Data part one, an *Employer* could include 'A, X7, X18.2, X18.3, Y(UK2) and Z . . .'.

**2.9.1.14 Option X20 – Key Performance Indicators (not used with Option X12)**

Option X20 Key Performance Indicators (KPIs) is for use in contracts where the Partnering Option X12 is not used. This Option reflects the trend towards the inclusion of KPIs in contracts and rewards *Contractors* for achieving an aspect of performance, for example, health and safety, time, programme.

Clause X20.1 requires an Incentive Schedule to be produced which identifies aspects of performance for which the *Contractor* is to be incentivised. The *Contractor* is required by clause X20.2 to report to the *Project Manager* his performance and forecast of final measurement against each KPI at the intervals stated in the Contract Data from the *starting date* until the Defects Certificate.

If at the performance assessment interval the *Contractor* is not on target to achieve the target set in the Incentive Schedule, he submits to the *Project Manager* his proposals for improving performance (clause X20.3). Note that there are no penalties for not achieving the target KPIs.

The *Contractor* is paid (clause X20.4) the amount stated in the Incentive Schedule if the target stated is achieved or improved upon. Payment of the amount due is when the target has been improved upon or achieved.

Clause X20.5 allows the *Employer* to add a KPI and associated payment to the Incentive Schedule. He is not allowed to delete or reduce a payment stated in the Incentive Schedule.

The layout for the Incentive Schedule will be similar to that contained in the Partnering Option X12.

### 2.9.1.15 Option Y(UK)2 – Part II of the Housing Grants, Construction and Regeneration Act 1996

Option Y(UK)2 was introduced to cover the introduction of the Housing Grants, Construction and Regeneration Act 1996 (HGCR) (Part II) for use with the ECC. This Option was prepared solely for use on contracts which are subject to the HGCR; that is, for contracts that fall under the definition of a construction contract under the Act.

The two principles contained in the Act are those related to payment and adjudication. If a contract is subject to the Act but Option Y(UK)2 was not chosen as part of the contract strategy, then the Scheme for Construction Contracts is read into the contract to account for the payment and adjudication changes required by the Act.

Option Y(UK)2 in ECC3 refers only to payment. This is because adjudication is now incorporated into Options W1 and W2, where Option W2 is chosen if the HGCR applies to the contract.

### 2.9.1.16 Option Y(UK)3 – the Contracts (Rights of Third Parties) Act 1999

Option Y(UK)3 covers the introduction of the Contracts (Rights of Third Parties) Act 1999 for use with the ECC when used in England and Wales. This law is covered by common law within Scotland and is therefore not required for contracts for which the *law of the contract* is the law of Scotland. The optional clause should be included in all contracts in England and Wales that come into existence on or after 11 May 2000.

The Act provides for the right of a person who is not a party to a contract (a 'third party') to enforce a term of the contract in certain circumstances. This is only when the contract expressly states so or it is the intention of the parties to do so. Option Y(UK)3 is intended to avoid giving third parties any rights inadvertently. If third parties are to be given rights in a particular contract, professional advice should be sought on matters such as

- which term of the contract the third party may enforce
- the identity of the third party
- when does the right of a third party arise
- a management procedure for being aware of a third party's right arising and advising the third party that it has arisen
- legal challenges (under section 2(1) of the Act) by a third party to prevent the parties to the contract from rescinding the contract or varying it so as to extinguish or alter his rights without his consent
- the establishment of dispute resolution procedures involving a third party exercising his rights under the Act and taking into account the requirements of the HGCR.

Y(UK)3 in ECC3 is included in the body of ECC3 in the secondary Option section. As such, the wording relating to the Act is already incorporated in the contract and therefore need not be included in Contract Data part one. Y(UK)3 in ECC3 does, however, require a Contract Data entry with regard to the term and the person/organisation who may enforce that term. It is suggested that, in most instances, the list would simply read 'None'.

### 2.9.1.17 Option Z – additional conditions of contract

Option Z introduces additional conditions of contract, with the caveat that any changes to the core clauses should be carefully considered. Because there are no cross-references within the ECC *conditions of contract*, a ready knowledge of the ECC is required before clauses are amended. Option Z should not be used to change the core clauses of the contract and to introduce onerous and adversarial terms and conditions. Option Z may be used for additional conditions such as confidentiality (particularly to do with the Freedom of Information Act 2000 for public sector bodies in the UK) or boundary conditions that are not included in the core clauses.

---

### Examples

Z2　The second sentence of clause 43.2 is deleted and replaced as follows: 'This period begins when the Defect is notified.'

Z3　Clause 60.1(4) is deleted and replaced with the following: 'The *Project Manager* gives an instruction to stop or not to start any work or change a Key Date except where the instruction relates to health and safety matters or is in relation to a *Contractor* default.'

Z4　The *Contractor* submits his valid tax invoice seven days after the date of the payment certificate. Where the *Contractor* does not submit his valid tax invoice within the time required

- the period within which payment is made and
- the time allowed in clause 91.4

are extended by the length of time from the date when the *Contractor* should have submitted his valid tax invoice to the date when he does submit it.

Z8　Clause 2 Equipment of the full Schedule of Cost Components is deleted and replaced with Clause 2 Equipment of the Shorter Schedule of Cost Components.

Note that these examples are examples only and do not represent recommendations by the authors. All changes to the *conditions of contract* should be vetted by the *Employer*'s legal representatives.

---

**Procuring an Engineering and Construction Contract**
ISBN 978-0-7277-5720-3

ICE Publishing: All rights reserved
doi: 10.1680/pecc.57203.069

# Appendix 3
# Audit Plan

## Section A The Audit Plan – generally

**A3.1 Introduction**

This plan has been produced to audit the Defined Costs for the *works* undertaken using the Engineering and Construction Contract Option C – Target Contract with Activity Schedule/Option D – Target Contract with Bill of Quantities/Option E – Cost Reimbursable Contract.

The Engineering and Construction Contract provides that the *Contractor* is paid the Price for Work Done to Date plus other amounts to be paid to the *Contractor* less amounts to be paid by or retained from the *Contractor* (clause 50.2).

Under Option C – Target Contract with Activity Schedule/Option D – Target Contract with Bill of Quantities/Option E – Cost Reimbursable Contract, the Price for Work Done to Date is

■ the total Defined Cost which the *Project Manager* forecasts will have been paid by the *Contractor* before the next assessment date plus the Fee (clause 11.2(29)).

This Audit Plan is designed to demonstrate that the assessment of the *Contractor*'s Defined Cost has been completed in accordance with the provisions of the *conditions of contract* and satisfies the requirements of third-party or external auditors.

The *Employer*, the *Project Manager* and the *Contractor* (called the Project Team for the purposes of this Audit Plan) are jointly committed to ensuring the control of Defined Cost on the *works*. This commitment is reinforced by their joint liability for any overrun of Defined Cost provided by the share mechanism in the *conditions of contract* (for Options C and D).

The Project Team will operate the procedures within this plan.

**A3.2 Basis of Audit Plan and procedures**

This procedure is devised on the basis that there is a joint commitment towards minimising costs due to the target cost and incentive schemes incorporated within the contract. With the joint incentive to minimise the cost, relatively uncomplicated audit procedures could be adopted. This Audit Plan avoids the need for auditing all costs and endeavours to minimise cost of implementing the procedures.

There is a need for a number of audit procedures to be established to meet the audit objectives and also to satisfy the requirements of third-party or external auditors.

The Audit Plan has been established to review administration and management procedures that are in place to manage the costs.

**A3.3 Main audit objectives**

The intention of the Audit Plan is to allow an auditor appointed from the Project Team to provide a monthly report based on a systematic analysis of the records of Defined Cost. This report is intended to provide the Project Team with the maximum confidence in the accuracy of the Defined Cost paid while employing the minimum practical resource to implement the procedures.

The Audit Plan is divided into seven key audit task sections

1  Subcontractors
2  People
3  Equipment
4  Plant and Materials
5  Charges
6  Manufacture and Fabrication
7  Design.

## A3.4 Reporting

The auditor, as a result of carrying out the audit, will generate a number of reports/comments. An example of a pro-forma 'Audit Comment Sheet' is given in Figure A3.1.

Reports will highlight anomalies as well as items that need further action; however, the auditor will, when requested, use his professional experience to provide comments/recommendations for consideration by others, which are intended to assist in the management of the *works*. Part of the pre-contract Section 2 – Subcontractors has been specifically drawn up for this purpose.

## A3.5 Audit report

Each section refers to the issue of an audit report which is closed out when it has been signed-off by a number of personnel.

Recommendations contained in audit reports will undoubtedly provide part direction to the audit tasks for subsequent period audits.

The Audit Plan will also need to be reviewed on a regular basis. Objectives and procedures will be scrutinised by the auditor and comments and observations will be welcomed from the Project Team. Due to this, a further audit section is included which will allow comments to be recorded and amendments to be made to the Audit Plan itself.

Comments and proposals to amend the Audit Plan will be discussed and agreed by the auditor and the *Employer*'s and *Contractor*'s commercial managers.

## A3.6 Goals

Each month the auditor will set the 'goals' of the Audit Plan for the next month. This will be based upon

- meeting the main audit objectives
- recommendations/comments arising from previous monthly audit reports
- total workload levels for the month.

The auditor will endeavour to work to a strict timetable in order to meet the audit objective and goals, and it is vital therefore that information from the Project Team is provided as rapidly as possible and without hindrance.

## A3.7 Amendments to Audit Plan

The Audit Plan is divided into eight sections. These are drafted to enable each section to be self-contained and therefore general information is included within each section.

The Audit Plan will be given a new revision letter upon amendment.

## A3.8 Distribution of Audit Plan

The Audit Plan will be distributed by the auditor to the Project Team, *Employer*'s and *Contractor*'s head offices and any third-party or external auditors at the request of the *Project Manager*.

## Section B Subcontractors

Subcontractors are as defined in clause 11.2(17) of the *conditions of contract*.

## B3.1 Audit objectives

The Subcontractor audit objectives are divided into (1) Pre-Award and Award and (2) Post-Award tasks to verify that:

**Figure A3.1** Audit Comment Sheet

Audit Comment Sheet

Contract title:

Project Reference No.:

*Employer*:
*Project Manager*:
*Supervisor*:
*Contractor*:

Details of Audit:
[*Description of audit to be carried out*]

Date of Audit:                              Auditor:
[*Insert Date*]                              [*Insert Name*]

Audit Findings:

Auditor _____                    Date _____

Copies to:

1 Pre-Award and Award
- subcontract tender documentation is accurate and
- contractually fair and reasonable
- competitive prices are obtained
- subcontracts are fairly awarded.

2 Post-Award
- subcontracts are properly administered
- interim payments made to subcontractors are fair and reasonable
- final accounts are completed.

### B3.1.1 Pre-Award and Award

When requested by the Project Team, the audit will be undertaken to

1 confirm compilation of proposed subcontract tender list,
2 verify:
- consistency in contractual terms between the various subcontract tenders issued for this project
- tenders are based on latest design and programme information
- terms and conditions are fair and reasonable, take cognisance of the main contract conditions and comply with current legislation regarding subcontracts
- reasonable price breakdowns have been requested
3 confirm that subcontract tender enquiries are competitively sought or determine reasons for single tender action
4 confirm that subcontract tenders are sent out and received back and opened at the same time
5 confirm that tenders are being sought at an appropriate time and that tender periods are reasonable
6 confirm that tender returns are compared reasonably and that tender qualifications are withdrawn as far as possible
7 confirm that tenders have been fairly assessed and that the most value-for-money option has generally been accepted (taking cognisance of any outstanding qualifications)
8 confirm that there is full disclosure of discounts
9 confirm that the tender to be accepted has a reasonable breakdown of prices to facilitate evaluation of variations
10 confirm that the tenders are awarded on the basis of latest information.

### B3.1.2 Post-Award

The audit will be undertaken to verify the following.

1 Changes to the subcontract are being instructed properly in a timely manner and procedures are in place to ensure that Subcontractors have the latest Works and Site Information issued by the *Project Manager*.
2 Payments made to Subcontractors are based upon the subcontract conditions of contract and where non-Engineering and Construction Subcontract (ECS) are based on:
- work done and/or materials on site
- subcontract rates and prices
- reasonable rates and prices (in respect of Compensation events where subcontract rates and prices do not apply).
3 All payments prepared by the *Contractor* are for work carried out or for materials supplied for this project and do not include Disallowed Cost.
4 Payments made to Subcontractors align with payments made to the *Contractor* for the same purpose.

### B3.2 Procedures
### B3.2.1 Generally

The auditor shall be supplied with or given access to all information necessary for compliance with the procedures contained within this plan.

1 Audit objectives B.1.1.1 and B.1.1.3.
2 For subcontract enquiries issued on this project, the auditor shall be provided with a signed-off list of tenderers.
3 Reasons for single tender actions shall be provided.

| | |
|---|---|
| **B3.2.2 Audit Objectives B.1.1.2, B.1.1.4 and B.1.1.5** | 1 For all subcontract enquiries to be issued on this project, the auditor shall be provided with a copy of the draft tender documentation. The auditor will undertake a spot check of the documentation within five working days of receipt. The auditor will verify that the tender documentation |

1 For all subcontract enquiries to be issued on this project, the auditor shall be provided with a copy of the draft tender documentation. The auditor will undertake a spot check of the documentation within five working days of receipt. The auditor will verify that the tender documentation

- includes latest drawings by reference to the Project Team's document control records
- is based upon the Engineering and Construction Subcontract or the *Contractor*'s standard form of subcontract
- is consistent for all enquiries, and main contractual terms are relevant and clear.

2 The review by the auditor shall not be allowed to delay/affect the issuing of the enquiry documents.

3 The auditor shall be provided with any subsequently issued documents detailing

- any amendments made to the draft tender documents or price
- amendments to be made but currently excluded and how they are to be incorporated in the future.

Note: the Project Team should carry out detailed checks on the accuracy and quality of the enquiry documentation.

**B3.2.3 Audit Objectives B.1.1.6 to B.1.1.10**

1 For all subcontract enquiries to be issued on this project, the auditor shall receive an initial summary of the tender offers, together with a summary of any qualifications included therein.

2 The Project Team shall assess the tenders in more detail, endeavour to remove qualifications and send the auditor a copy of their recommendation together with their reconciliation of prices and qualifications.

3 Within five working days of receipt, providing this date is before the date required for award, the auditor may review documents and provide any comments on the recommendation to the Project Team.

**B3.2.4 Audit Objective B.1.2.1**

1 When requested by the auditor, the Project Team shall provide a schedule listing all instructions issued to Subcontractors.

2 The auditor will carry out random checks by comparison to the Project Team's document control records.

**B3.2.5 Audit Objectives B.1.2.2 and B.1.2.3**

1 The auditor will carry out spot checks on payment certificates. When requested by the auditor, the Project Team shall provide a copy of the latest payment certificate together with a breakdown of the latest or previous valuations as may be necessary.

2 The auditor may request further support documentation such as

- site notes/measurements
- measurements from drawings/final accounts
- build-ups for new rates
- copies of subcontract documentation
- delivery tickets
- schedule of Disallowed Costs.

3 The auditor will carry out spot checks to verify

- rates or prices used are contract rates or prices or compatible therewith
- work has been executed in relation to this project
- the value included for materials relates to materials on site for this project
- the value of compensation events is identified
- consideration has been made for Disallowed Costs.

4 On completion of the draft final account, the auditor may review the final account and carry out spot checks as referred to above.

**B3.2.6 Audit Objectives B.1.2.4**

1 The auditor shall carry out spot checks that payments made to *Contractor* in respect of Subcontractors have also been made to the Subcontractors.

**B3.3 Audit records**
**B3.3.1 Generally**

Full audit records of Subcontractors will be kept and filed in accordance with this procedure.

**B3.3.2 Monthly Subcontractor Audit Report**

The auditor will issue a Subcontractor Audit Report to the Project Team on a monthly basis. It will identify positive findings as well as items needing further action and consideration.

The Subcontractor Audit Report will not be deemed completed until it has been signed off by

- the auditor
- the *Project Manager* or his authorised representative and
- the *Contractor* or his authorised representative.

**B3.3.3 Subcontractor Record of Audit**

The Record of Audit is prepared by the auditor to record all tasks undertaken since the cut-off date of the previous Subcontractor Audit Report. It is retained by the auditor for review by the Project Team and third-party or external auditors.

The Record of Audit will be the backup document to the Subcontractor Audit Report.

**B3.3.4 Filing**

A file will be established entitled '[Project Name] – Subcontractor Audit Records'.

**B3.3.5 Time-scales**

The auditor will initially carry out an audit for the month of [month] [year]. Subsequent audits will be carried out on a monthly basis.

## *Section C People*

Defined Cost relating to People is described in section 1 of the Schedule of Cost Components.

**C3.1 Audit objectives**

Examination of individual payslips of persons who are paid in full (or in part) on a time basis to verify

- payslips tie up with a weekly payroll summary provided by the *Contractor*
- rates of pay are in accordance with conditions of employment
- payslips relate to labour allocation/timesheets
- payments relate to work executed on this project.

**C3.2 Audit records**
**C3.2.1 Generally**

- Full audit records of People will be kept and filed in accordance with this procedure.
- It is anticipated that the People 'trend' charts as well as the 'People Audit Reports' will be of use both to the auditor and the Project Team.

**C3.2.2 Monthly People Audit Report**

The auditor will issue a People Audit Report to the Project Team on a monthly basis. It will identify positive findings as well as items needing further action and consideration.

The People Audit Report will not be deemed completed until it has been signed off by

- the auditor
- the *Project Manager* or his authorised representative and
- the *Contractor* or his authorised representative.

**C3.2.3 People Record of Audit**

1 The Record of Audit is prepared by the auditor to record all tasks undertaken since the cut-off date of the previous People Audit Report. It is retained by the auditor for review by the Project Team and third-party or external auditors.
2 The Record of Audit will summarise the records examined during the period and it will note
   - calendar period
   - basis of selection
   - names of People records audited
   - payroll numbers
   - grade of labour and work location
   - if allocation sheets/timesheet and swipe card records are consistent
   - if payslips tie up with weekly payroll records.

3 The People Record of Audit will be signed off by the auditor as a true record and forms the basis of the Monthly People Audit Report.

**C3.2.4 Trend tables**

The trend table(s) which will include the following will supplement the Record of Audit

- wage bill for current and previous periods
- average labour cost per person per month for this period and previous periods
- number of persons employed during this period and previous periods.

**C3.2.5 Filing**

A file will be established entitled '[Project Name] – People Audit Records'.

**C3.2.6 Time-scales**

The auditor will initially carry out an Audit for the month of [month and year]. Subsequent audits will be carried out on a monthly basis.

**C3.3 Information to be provided by the Project Team**

In order to meet the audit objectives, the Project Team will be required to provide information including, but not limited to

- weekly payroll build-up
- payslips
- allocation sheets
- productivity records
- gradings of staff
- agreed rates of pay and basis of agreement
- information from bank statements.

## Section D Equipment

For the purposes of this procedure, Equipment is as defined in clause 11.2(7) of the *conditions of contract*. Defined Cost relating to Equipment is described in section 2 of the *Schedule of Cost Components*.

**D3.1 Audit objectives**

The audit objectives are divided into two main areas to verify the following.

- Pre-order: orders or purchases are based upon accurate documentation and competitive prices.
- Post-order: Equipment orders are properly administered and payments made relate to usage on this project.

**D3.1.1 Pre-order**

1 Confirm competitive prices have been obtained or used and where plant is hired or purchased from other parts of the *Contractor* group of companies confirm that prices are reasonable.
2 Confirm site indents estimate the periods of hire and that indents have been signed off.

**D3.1.2 Post-order**

1 Confirm prices for extra period of hire are generally obtained as set out in the pre-order section or based upon prices previously agreed and obtained competitively.
2 Verify payments made are based upon plant hired or purchased, at contract rates and prices, and also relate to plant charges arising after [date].
3 Verify plant deliveries are checked and align with orders.
4 Check a procedure is in place to ensure plant is off-hired in a timely manner.
5 Verify payments made for plant align with payments made to *Contractor* and that payments are made at the appropriate time and that any Disallowed Costs have been taken into account.
6 Check overall capital expenditure costs allow for a residual value of purchased plant.

**D3.2 Procedures**
**D3.2.1 Audit Objectives D.1.1.1, D.1.2.1 and D.1.2.6**

- The auditor will review a list of Equipment orders placed.
- The auditor will select orders and view documentation and resulting reports and recommendations, spot check that competitive prices have been received and that costs incurred relate to this project.

- Where plant is hired or purchased from a company within the same group of companies as the *Contractor*, the auditor shall be provided with details of competitive rates and prices and that hire rates for specialist Equipment make due consideration for working life/residual value.

**D3.2.2 Audit Objective D.1.1.2**

- The auditor will select orders at random and view requisitions relating thereto.
- The auditor will carry out spot checks to verify
  - requisitions have been signed off in accordance with the relevant procedure
  - requisitions and Equipment orders identify the estimated period of hire
  - orders are sanctioned by the delegated person within the Project Team.

**D3.2.3 Audit Objectives D.1.2.1, D.1.2.2 and D.1.2.3**

- The auditor will select orders at random and review relevant delivery notes to ascertain that Equipment has been checked and received and is intended for use on this project.
- The auditor will select orders at random to carry out spot checks. These checks shall include checking that payments are compatible with deliveries, periods of hire and invoices received. Also that the correct rates have been used for calculation of payment due, all discounts are disclosed and that periods of additional hire are backed up by revised requisitions based on site assessments.

**D3.2.4 Audit Objective D.1.2.4**

- The auditor will carry out spot checks to determine that procedures are in place to monitor the use of/need for Equipment and that Equipment is being returned in a timely manner.
- The auditor will check requisition forms and compare these with monitoring procedures.

**D3.2.5 Audit Objective D.1.2.5**

- The auditor shall spot check that amounts certified by the Project Team align with total payments made in respect of Equipment.

## D3.3 Audit records
**D3.3.1 Generally**

Full audit records of plant will be kept and filed in accordance with this procedure.

**D3.3.2 Equipment Audit Report**

The auditor will issue an Equipment Audit Report to the Project Team on a monthly basis. It will identify positive findings as well as items needing further action and consideration.

The Equipment Audit Report will not be deemed completed until it has been signed off by

- the auditor
- the *Project Manager* or his authorised representative
- the *Contractor* or his authorised representative.

**D3.3.3 Equipment Record of Audit**

- The Record of Audit is prepared by the auditor to record all tasks undertaken since the cut-off date of the previous Equipment Audit Report. It is retained by the auditor for review by the Project Team and third-party or external auditors.
- The Record of Audit will summarise the records examined during the period and it will note
  - calendar period
  - basis of selection
  - items of Equipment audited
  - Equipment identification numbers
  - if requisitions correlate with orders and invoices
  - if procedures for off-hiring Equipment have been operated in the period.
- The Equipment Record of Audit will be signed off by the auditor as a true record and forms the basis of the Monthly Equipment Audit Report.

**D3.3.4 Filing**

A file will be established entitled '[project name] – Equipment Audit Records'.

**D3.3.5 Time-scales**

The auditor will initially carry out an Audit for the month of [month] [year]. Subsequent audits will be carried out on a monthly basis.

**D3.4 Information to be provided by the Project Team**

In order to meet these objectives, the Project Team will be required to provide information including, but not limited to

- list of orders raised and estimated values
- summary of payments made in the period
- quotations from suppliers
- access to invoices and delivery note records
- schedule of total plant on site
- assessments/recommendations carried out by Project Team prior to placing orders
- site assessment of schedules of plant.

## Section E Plant and Materials

For the purposes of this procedure, Plant and Materials are as defined in clause 11.2(12) of the *conditions of contract*. Defined Cost relating to Plant and Materials is described in section 3 of the Schedule of Cost Components.

**E3.1 Audit objectives**

The audit objectives are divided into two main areas to verify the following.

- Pre-order: orders are based upon accurate documentation and competitive prices are obtained.
- Post-order: that material orders are properly administered and that payments are accurate and timely.

**E3.1.1 Pre-order**

1 Confirm competitive prices are obtained.
2 Confirm orders are, where applicable, based upon a reasonable assessment of quantities using the latest Works and Site Information.
3 Confirm a maximum order value based upon quantities is ascertained and recorded.
4 Confirm orders detail payment periods and terms.
5 Confirm the final location of materials can be identified from the order and requisition.
6 Confirm details on orders match those on the requisition raising the order.

**E3.1.2 Post-order**

1 Verify payments made are based upon materials delivered and agreed rates and payments exclude any Disallowed Cost.
2 Verify material delivery notes are being accurately checked.
3 Verify total delivery of materials corresponds with orders.
4 Confirm payments made for materials align with payments made to *Contractor* for same purpose and payments are made at the appropriate time.

**E3.2 Procedures**
**E3.2.1 Audit Objectives E.1.1.1 and E.1.1.5**

- The auditor will select orders and view indents and order forms and verify competitive prices have been received and that the location for materials can be ascertained.
- The basis of selection will be recorded.

**E3.2.2 Audit Objective E.1.1.2**

- The auditor will select an order at random and review quantities against drawing numbers referred to in the order.
- The auditor will carry out spot checks to determine that
  - latest drawings have been used
  - quantities have been used as the basis for enquiries and can be ascertained from the drawings.

Note: The auditor will not carry out detailed checks on the quantities and orders as the Project Team should carry this out.

**E3.2.3 Audit Objective E.1.1.3**

- The auditor will carry out spot checks that maximum prices recorded coincide with tenders received.

**E3.2.4 Audit Objective E.1.1.4**

- The auditor will carry out spot checks on selected enquiries and orders to determine whether payment terms/payment periods are recorded and reasonable.

| | |
|---|---|
| **E3.2.5 Audit Objective E.1.1.5** | ■ The auditor will carry out spot checks on selected orders to determine whether requisition details match those on orders. |
| **E3.2.6 Audit Objectives E.1.2.1, E.1.2.2 and E.1.2.3** | ■ The auditor will select orders and request access to delivery notes and invoices and will carry out spot checks using selected delivery notes to ascertain that materials have been received and checked and that a reconciliation has been carried out against the order.<br>■ The auditor will select orders to carry out spot checks that payments made are consistent with actual deliveries made and that correct rates have been used for this purpose, and that all discounts are disclosed. |
| **E3.2.7 Audit Objective E.1.2.4** | ■ The auditor shall carry out spot checks that amounts certified by the Project Team for payment align with total payments made. |

**E3.3 Audit records**
**E3.3.1 Generally**

Full audit records of Plant and Materials will be kept and filed in accordance with this procedure.

**E3.3.2 Plant and Materials Audit Report**

The auditor will issue a Plant and Materials Audit Report to the Project Team on a monthly basis. It will identify positive findings as well as items needing further action and consideration.

The Plant and Materials Audit Report will not be deemed completed until it has been signed off by

■ the auditor
■ the *Project Manager* or his authorised representative
■ the *Contractor* or his authorised representative.

**E3.3.3 Plant and Materials Record of Audit**

■ The Record of Audit is prepared by the auditor to record all tasks undertaken since the cut-off date of the previous Plant and Materials Audit Report. It is retained by the auditor for review by the Project Team and third-party or external auditors.
■ The Record of Audit will summarise the records examined during the period and it will note
  ■ calendar period
  ■ basis of selection
  ■ items of Plant and Materials audited
  ■ if requisitions correlate with orders and invoices.
■ The Plant and Materials Record of Audit will be signed off by the auditor as a true record and forms the basis of the Monthly Plant and Materials Audit Report.

**E3.3.4 Filing**

A file will be established entitled '[project name] – Plant and Materials Audit Records'.

**E3.3.5 Time-scales**

The auditor will initially carry out an Audit for the month of [month and year]. Subsequent audits will be carried out on a monthly basis.

**E3.4 Information to be provided by the Project Team**

In order to meet these objectives, the Project Team will be required to provide information including, but not limited to

■ Plant and Materials requisitions
■ list of orders raised and values of orders
■ register of delivery notes for the period and access to invoices and records
■ invoices received.

## Section F Charges

For the purposes of this procedure Defined Cost relating to Charges is described in section 4 of the Schedule of Cost Components.

## F3.1 Audit Objectives

### F3.1.1 Random examination of payments

This is done to verify

1. a comprehensive build-up is available
2. where applicable, costs are based upon quotes/invoices/scale charges/fair rates and that costs exclude Disallowed Cost
3. costs need to be incurred to carry out the *works*
4. costs relate to this project.

### F3.1.2 Spot checks on overall costs

These are to

1. compare site costing records against actual pay records sent from the *Contractor*
2. record cost trends.

## F3.2 Procedures

### F3.2.1 Generally

- Due to the nature of the contract, it is impractical to check all records of Charges and the audit must therefore necessarily be selective.

### F3.2.2 Audit Objective F1.1

- The auditor will select one or more payments in the audit period.
- The auditor shall determine and record the basis of selection.
- The auditor inspects the *Contractor*'s build-up to the payment to ensure it is comprehensive and excludes any Disallowed Cost.
- The auditor also selects individual costs in the build-ups. The *Contractor* will provide evidence of the basis of order and payment and to ascertain if prices are based on competitive rates/scale charges or fair rates.

### F3.2.3 Audit Objective F1.2

- The total costs for Charges are spot checked against actual records sent from the *Contractor*.
- A trend table is produced comparing expenditure for different periods.
- The *Contractor* shall on a monthly basis break down the expenditure into agreed heads of charge. The auditor will check costs and enter them into a trend table, comparing expenditure on a monthly basis.

## F3.3 Audit records

### F3.3.1 Generally

- Full audit records of Charges will be kept and filed in accordance with this procedure.
- It is anticipated that the trend tables will be of use to both the auditor and the Project Team.

### F3.3.2 Charges Audit Report

The auditor will issue a Charges Audit Report to the Project Team on a monthly basis. It will identify positive findings as well as items needing further action and consideration.

The Charges Audit Report will not be deemed completed until it has been signed off by

- the auditor
- the *Project Manager* or his authorised representative
- the *Contractor* or his authorised representative.

### F3.3.3 Charges Record of Audit

- The Record of Audit is prepared by the auditor to record all tasks undertaken since the cut-off date of the previous Charges Audit Report. It is retained by the auditor for review by the Project Team and third-party or external auditors.
- The Record of Audit will summarise the records examined during the period and it will note
  - calendar period
  - basis of selection
  - if payments correlate with orders and invoices.
- The Charges Record of Audit will be signed off by the auditor as a true record and forms the basis of the Monthly Charges Audit Report.

### F3.3.4 Trend tables

The trend table(s) which will include the following will supplement the Record of Audit.

- Total cost for current and previous periods
- Total cost for current and previous periods in agreed head of charge.

**F3.3.5 Filing**  A file will be established entitled '[project name] – Charges Audit Records'.

**F3.3.6 Time-scales**  The auditor will initially carry out an audit for the month of [month] [year]. Subsequent audits will be carried out on a monthly basis.

**F3.4 Information to be provided by the Project Team**  In order to meet these objectives, the Project Team will be required to provide information including, but not limited to

- monthly build-up of costs broken down into heads of charge
- agreed rates of pay/quotations/invoices, etc.

## Section G Manufacture and Fabrication

For the purposes of this procedure, Defined Cost relating to Manufacture and Fabrication is described in section 5 of the Schedule of Cost Components.

**G3.1 Audit objectives**
**G3.1.1 Examination of build-up to payments**  This is done to confirm

1  value of Equipment and Plant and Materials is in agreement with contract values
2  calculation of price adjustment factor is correct
3  date for completion of Manufacture and Fabrication of Equipment and Plant and Materials is agreed with designated member of the Project Team.

**G3.1.2 Visit site of Manufacture and Fabrication**  This is done to

1  confirm items of Equipment and Plant and Materials included in payment are complete
2  verify Equipment and Plant and Materials are marked in accordance with the procedures described in the Works Information (clause 71.1).

**G3.2 Procedures**
**G3.2.1 Generally**
- The auditor shall be supplied with or given access to all information necessary for compliance with the procedures contained within this plan.
- The auditor shall be allowed access to all sites where Equipment and Plant and Materials are stored outside the Working Areas, having given due notice of his intention to make an inspection.

**G3.2.2 Audit Objective G1.1**
- The auditor shall be provided with calculations detailing the build-up to sums included in payments for Manufacture and Fabrication outside the Working Areas.
- The auditor shall be provided with a notification confirming the date of completion of Manufacture and Fabrication of Equipment and Plant and Materials.

**G3.2.3 Audit Objective G1.2**
- The auditor shall visit the site of Manufacture and Fabrication outside the Working Areas to view completed Equipment and Plant and Materials, and check items are marked in accordance with the procedures described in the Works Information.

**G3.3 Audit records**
**G3.3.1 Generally**  Full audit records of Manufacture and Fabrication will be kept and filed in accordance with this procedure.

**G3.3.2 Charges Audit Report**  The auditor will issue a Manufacture and Fabrication Audit Report to the Project Team on a monthly basis. It will identify positive findings as well as items needing further action and consideration.

The Manufacture and Fabrication Audit Report will not be deemed completed until it has been signed off by

- the auditor
- the *Project Manager* or his authorised representative
- the *Contractor* or his authorised representative.

**G3.3.3 Charges Record of Audit**

- The Record of Audit is prepared by the auditor to record all tasks undertaken since the cut-off date of the previous Manufacture and Fabrication Audit Report. It is retained by the auditor for review by the Project Team and third-party or external auditors.
- The Record of Audit will summarise the records examined during the period and it will note
  - calendar period
  - basis of selection
  - if payments correlate with orders and invoices.
- The Manufacture and Fabrication Record of Audit will be signed off by the auditor as a true record and forms the basis of the Monthly Manufacture and Fabrication Audit Report.

**G3.3.4 Filing**

A file will be established entitled '[project name] – Manufacture and Fabrication Audit Records'.

**G3.3.5 Time-scales**

The auditor will initially carry out an audit for the month of [month] [year]. Subsequent audits will be carried out on a monthly basis.

## Section H Design

For the purposes of this procedure, Defined Cost relating to design is described in section 6 of the Schedule of Cost Components.

**H3.1 Audit objectives**
**H3.1.1 Examination of payslips (time basis)**

Examination of individual payslips of persons who are paid in full (or in part) on a time basis to verify

1 payslips tie up with a weekly payroll summary provided by the *Contractor*
2 rates of pay are in accordance with conditions of employment
3 payslips relate to labour allocation/timesheets
4 payments relate to Work executed on this project.

**H3.1.2 Examination of payslips (productivity basis)**

Examination of individual payslips of persons paid in full (or in part) on a productivity basis to verify

1 payslips tie up with weekly payroll summary provided by the *Contractor*
2 productivity payments tie up with productivity records.

**H3.1.3 Examination of consistency**

Examination to confirm consistency between labour allocation sheets and swipe card checking on/off system.

**H3.1.4 Spot checks on labour costs**

Spot checks on overall labour costs to

1 compare site costing records against actual pay records provided by the *Contractor*
2 calculate average labour cost per person-week for the period
3 calculate average hourly cost for the period
4 produce trend tables to show changes in costs and workforce levels
5 ensure that Disallowed Costs have not been included in payments.

**H3.2 Procedures**
**H3.2.1 Generally**

Due to the nature of the contract, it is impractical to check the pay records of all design personnel employed by the *Contractor* on the project. The audit, therefore, must be based on a selective cross-section that is representative of the Defined Cost incurred.

**H3.2.2 Audit Objectives H1.1 and H1.2**

- The auditor will determine the time periods to be audited and select names from the list of design personnel employed on the Project.
- The auditor shall determine and record the basis of selection.
- The auditor shall inspect their allocation sheets/timesheets to ensure there are no discrepancies. The rates of pay shall be checked against a list of rates of pay and

sundry information determined from the conditions of employment provided by the *Contractor*.

■ Any overtime payments or travelling and subsistence payments are also checked for accuracy against the basis for calculation of such payment provided by the *Contractor*.

■ Payslips are also compared with weekly payroll summaries.

### H3.2.3 Audit Objective H1.3

■ The auditor determines the time periods to be audited and selects names from the list of design personnel employed on the project.

■ The auditor shall determine or record the basis of selection.

■ The auditor inspects allocation sheets they have been signed off by an authorised representative of the Project Team.

### H3.2.4 Audit Objective H1.4

The auditor determines the time period to be audited and how regularly calculations are to be made and uses the total Defined Cost of Design to

■ check the actual pay records supplied by the *Contractor*

■ calculate average labour cost per person, per week during the period and enter onto a 'trend' chart

■ calculate the average hourly cost during the period and enter onto a 'trend' chart

■ calculate the total number of persons employed and enter onto a 'trend' chart

■ check to ensure any Disallowed Costs have not been paid.

## H3.3 Audit records
### H3.3.1 Generally

■ Full audit records of design will be kept and filed in accordance with this procedure.

■ It is anticipated that the design 'trend' charts as well as the 'Design Audit Reports' will be of use both to the auditor and the Project Team.

### H3.3.2 Monthly Design Audit Report

The auditor will issue a Design Audit Report to the Project Team on a monthly basis. It will identify positive findings as well as items needing further action and consideration.

The Design Audit Report will not be deemed completed until it has been signed off by

■ the auditor

■ the *Project Manager* or his authorised representative

■ the *Contractor* or his authorised representative.

### H3.3.3 Design Record of Audit

■ The Record of Audit is prepared by the auditor to record all tasks undertaken since the cut-off date of the previous Design Audit Report. It is retained by the auditor for review by the Project Team and third-party or external auditors.

■ The Record of Audit will summarise the records examined during the period and it will note
  ■ calendar period
  ■ basis of selection
  ■ names of design personnel records audited
  ■ payroll numbers
  ■ grade of labour and work location
  ■ if payslips tie up with weekly payroll records.

■ The Design Record of Audit will be signed off by the auditor as a true record and forms the basis of the Monthly Design Audit Report.

### H3.3.4 Trend tables

The trend table(s) which will include the following will supplement the Record of Audit.

■ Wage bill for current and previous periods.

■ Average labour cost per person per month for this period and previous periods.

■ Number of persons employed during this period and previous periods.

### H3.3.5 Filing

A file will be established entitled '[project name] – Design Audit Records'.

### H3.3.6 Time-scales

The auditor will initially carry out an audit for the month of [month] [year]. Subsequent audits will be carried out on a monthly basis.

**H3.4 Information to be provided by the Project Team**

In order to meet the audit objectives, the Project Team will be required to provide information including, but not limited to

- weekly payroll build-up
- payslips
- allocation sheets
- productivity records
- grading of staff
- agreed rates of pay and basis of agreement
- information from bank statements.

**Procuring an Engineering and Construction Contract**
ISBN 978-0-7277-5720-3

doi: 10.1680/pecc.57203.085

# Chapter 3
# Completing the Contract Data

**Synopsis**　　　　This chapter gives guidance on

- how to choose a main Option

- how to choose secondary Options

- choosing optional statements in the Contract Data

- where to position the optional statements in the Contract Data

- how to complete each statement in the Contract Data.

## 3.1 Introduction

The Contract Data should not be copied from previous contracts but should be considered carefully for each new contract that is drafted, taking into account the specific circumstances of the project.

The following information is designed to assist you in completing the Contract Data, by asking the questions that you will need to answer before you can complete the Contract Data.

The Contract Data represents the *Employer*'s contract strategy included in his invitation to tender documentation and the *Contractor*'s tender, and therefore forms the heart of an ECC contract. The same questions should be asked for each and every contract to make sure that appropriate choices are made and to make sure that the contract strategy suits the particular project.

Note that the chapter refers to the Contract Data in the ECC and therefore it refers to both Contract Data part one by the *Employer* and Contract Data part two by the *Contractor*. The information can just as easily be used by the *Contractor* for the Engineering and Construction Subcontract (ECS), where Contract Data part one is by the *Contractor* and Contract Data part two is by the Subcontractor.

## 3.2 Structure of the chapter

The columns of the table in section 3.4.1 below contain the following information.

- Column A: a question that needs to be answered.
- Column B: comments indicating how the Contract Data is affected by your answer.
- Column C: a replication of the Contract Data item that the question affects, so that you know which part of the Contract Data to complete.

The structure of the chapter is as follows.

1   The first part of the chapter asks questions about how you want to pay the *Contractor* (or the Subcontractor for an ECS contract) during the period of the contract, guiding you to a choice of main Option.
2   The second part of the chapter asks further questions about the contract strategy for the particular project, guiding you in your choice of secondary Options.
3   The third part of the chapter continues with questions about your chosen contract strategy, guiding you in choices about the optional statements that appear towards the end of the Contract Data.
4   The fourth part of the chapter describes each entry in Contract Data part one by the *Employer*, referring back to the contract strategy chosen through answering earlier questions in parts one, two and three above.
5   The fifth part of the chapter describes each entry in Contract Data part two by the *Contractor* (or the Subcontractor in the ECS), referring back to the contract strategy chosen through answering earlier questions.

| Secondary Option clauses (ECC3) | | | |
|---|---|---|---|
| X1 | Price adjustment for inflation | X15 | Limitation of the *Contractor*'s liability for his design to reasonable skill and care |
| X2 | Changes in the law | | |
| X3 | Multiple currencies | | |
| X4 | Parent company guarantee | X16 | Retention |
| X5 | Sectional Completion | X17 | Low-performance damages |
| X6 | Bonus for early Completion | X18 | Limitation of liability |
| X7 | Delay damages | X20 | Key Performance Indicators |
| X12 | Partnering | Y(UK)2 | The Housing Grants, Construction and Regeneration Act 1996 |
| X13 | Performance bond | | |
| X14 | Advanced payment to the *Contractor* | Y(UK)3 | The Contracts (Rights of Third Parties) Act 1999 |
| | | Z | Additional *conditions of contract* |

### 3.3 Part one of this chapter: choosing the main Option

This part of the chapter asks questions about how the *Employer* wants to pay the *Contractor* during the period of the contract, guiding you to a choice of main Option. The questions apply equally to the *Contractor*'s choices in the Engineering and Construction Subcontract (ECS), although those choices are normally more restricted.

#### 3.3.1 Contract Data by the *Employer*

| | Column A<br>Question | Column B<br>Assistance | Column C<br>Contract Data item |
|---|---|---|---|
| 1 | What is the status of the Works Information? How complete is it and how many changes are expected? | **Discussion**<br>The status of the Works Information affects the choice of main Option. The more complete the Works Information, the more suitable it will be to choose a lump sum contract as the payment mechanism. Other factors affecting the choice of main Option could be the perceived requirement to incentivise the *Contractor*.<br>**Action**<br>Choose the main Option that best matches the status and quality of the Works Information and best reflects the project objectives. | • The *conditions of contract* are the core clauses and the clauses for main Option .................... dispute resolution Option .................... and secondary Options .................... of the NEC3 Engineering and Construction Contract (June 2006 and September 2011). |
| 2 | How do you want to pay the *Contractor*; that is, using a *bill of quantities*, *activity schedule* or Defined Cost? | **Discussion**<br>The main Options A to F give a choice of payment mechanism. Part of the decision to use main Option A or B, or C or D, is a choice whether to pay according to a *bill of quantities* (remeasurement) or an *activity schedule* (milestone/stage payments).<br>**Action**<br>Options A and C use an *activity schedule* to build up the contract value and the target cost respectively. Options B and D use a *bill of quantities* to build up the contract value and the target cost respectively. Options C, D and E all use Defined Cost plus Fee as defined to pay the *Contractor* during the period of the contract, irrespective in the case of C and D, of the method of arriving at the target cost. | See Contract Data entries for Options A, B, C, D, E or F in both Contract Data part one and Contract Data part two. |

## 3.4 Part two of this chapter: choosing the secondary Options

This part of the chapter asks further questions about the *Employer*'s contract strategy, guiding you to choices about the secondary Options.

Most of the secondary Options could increase the cost of the project to the *Employer*. For example, it costs the *Contractor* to provide a performance bond or retention, and he is likely to pass that cost on to the *Employer* in some way. By offering these contract strategies as options rather than as standard, as other contracts do, the *Employer* can decide the individual risks and complexities of the contract and what secondary Options are required for a particular contract.

### 3.4.1 Contract Data by the *Employer*

| | Column A<br>Question | Column B<br>Assistance | Column C<br>Contract Data item |
|---|---|---|---|
| 3 | Do you want a performance bond from the *Contractor*?<br><br>YES  choose Option X13<br>NO  do nothing | **Discussion**<br>A performance bond provides a backup source of resource if the *Contractor* fails to complete the contract.<br><br>**Action**<br>If you want a performance bond, choose Option G, and include the format of the bond into the Works Information. | See Contract Data entries for Option X13 below |
| 4 | Do you want a parent company guarantee from the *Contractor*?<br><br>YES  choose Option X4<br>NO  do nothing | **Discussion**<br>A parent company guarantee is used for a similar purpose to the performance bond, except that usually physical resources, such as labour, is provided rather than solely financial resources. In addition, a parent company guarantee can only be provided if the *Contractor* has a parent company. If you want this kind of guarantee, either a performance bond or a parent company guarantee should be chosen rather than both.<br><br>**Action**<br>If you want a parent company guarantee rather than a performance bond, choose Option X4 and include the format of the bond in the Works Information. | There are no Contract Data entries for Option X4 |
| 5 | Does the project involve a substantial capital outlay by the *Contractor* and is the *Employer* willing to finance it?<br><br>YES  choose Option X14<br>NO  do nothing | **Discussion**<br>If a large item of Equipment is required to be purchased before the work starts, the *Employer* may choose to make an advance payment to the *Contractor* to assist his cash flow. The *Employer* may also require a bond from the *Contractor*, which, of course, will cost the *Contractor* money to obtain.<br><br>**Action**<br>If you want to advance the *Contractor* payments, then choose Option X14. If a bond is required, then include the format of the bond in the Works Information. An alternative to choosing Option X14 is to include an item in the *activity schedule* or *bill of quantities* for the purchase of the item. | See Contract Data entries for Option X14 below |
| 6 | Where payment is made to the *Contractor* in more than one currency, does the *Employer* carry the risk of the *exchange rate*?<br><br>YES  choose Option X3<br>NO  do nothing | **Discussion**<br>The effect of this Option is that the *Contractor* is protected from fluctuations in currency exchange rates where he is paid in foreign currencies for parts of the *works*.<br><br>**Action**<br>For Options A and B only, where the *Employer* wishes to take the risk of *exchange rates*, choose Option X3. Provision is made for multiple currencies within Options C, D, E and F, and therefore Option X3 should not be used with Options C, D, E and F. | See Contract Data entries for Option X3 below |

|  | Column A<br>Question | Column B<br>Assistance | Column C<br>Contract Data item |
|---|---|---|---|
| 7 | Do you want to take over parts of the *works* as they are completed?<br>YES choose Option X5<br>NO do nothing | **Discussion**<br>Sectional completion should be chosen if you want parts of the *works* to be completed before the whole of the *works*, where the parts of the works are key milestone dates leading up to Completion of the whole of the *works*. Sectional completion facilitates delay damages for key dates. The *Employer* must take over those parts of the *works* that have been completed, however, so unless you want to take over each section of the *works* as it is completed, there is no point in choosing sectional completion just because you can deduct delay damages if the *Contractor* is late in completing that section.<br>**Action**<br>If you want sectional completion – that is, if you want to take over *sections* of the *works* as they are completed – choose Option X5. | See Contract Data entries for Option X5 below. |
| 8 | Is the *Contractor* required to design part of or the whole of the *works*?<br>YES choose Option X15<br>NO do nothing | **Discussion**<br>The standard for the *Contractor*'s design liability is fitness for purpose. It is often difficult for a designer to insure for more than reasonable skill and care. Option X15 reduces the design liability to that of reasonable skill and care. It should be remembered that even with reasonable skill and care there is still a requirement for the design to be fit for its purpose, e.g. a contractor-designed warehouse floor should be suitable for fork-lift trucks to run over them.<br>**Action**<br>Choose Option X15 to reduce the *Contractor*'s liability for Defects due to his design from fitness for purpose to reasonable skill and care. | There are no Contract Data entries for Option X15. |
| 9 | Does the *Employer* want to take the risk for inflation?<br>YES choose Option X1<br>NO do nothing | **Discussion**<br>Without Option X1, the contract is a firm price and the *Contractor* takes the risk for inflationary increases in the costs of labour, plant and materials (for main Options A, B, C and D only). This would generally be applicable for a contract longer than approximately one year and/or during periods of high inflation.<br>**Action**<br>Choose Option X1 if the *Employer* wants to accept the risk for inflationary changes. | See Contract Data entries for Option X1 below. |
| 10 | Does the *Employer* want to retain a fund at the end of the contract to ensure the correction of Defects?<br>YES choose Option X16<br>NO do nothing | **Discussion**<br>The purpose of retention is to enable the *Employer* to retain a proportion of the amount paid to the *Contractor* as security for the correction of Defects after Completion and as an additional motivation for the *Contractor* to complete the *works*. The procedure in the ECC is slightly different from traditional contracts. In traditional contracts a retention percentage, typically 2.5%, is withheld on each and every interim payment from the outset of the contract. The ECC has a '*retention-free amount*'. The guidance notes suggest that this could be set at say 70% of the contract sum, the idea being that the *retention-free amount* aids the *Contractor*'s cash flow in the early part of the contract when he most needs it. Once the *retention-free amount* has been reached a higher *retention percentage* than normal is suggested. The net effect is the same except that the retention fund is built up at the latter part of the contract.<br>**Action**<br>If you think you will need added security, then choosing Option X16 is a method of achieving this. | See Contract Data entries for Option X16 below. |

| | Column A<br>Question | Column B<br>Assistance | Column C<br>Contract Data item |
|---|---|---|---|
| 11 | Will early Completion benefit the *Employer*?<br><br>YES  choose Option X6<br>NO   do nothing | **Discussion**<br>Option X6 can be used to motivate the *Contractor* to achieve early Completion by providing the *Contractor* with an early Completion bonus payment. This can be used where such early Completion would benefit the *Employer*, for example, opening of a retail outlet earlier completion = early income from outlet. Some *Employers* may use the potential income from an earlier opening date to calculate the bonus per day for early Completion.<br><br>**Action**<br>Where early Completion would benefit the *Employer*, choose Option X6. | See Contract Data entries for Option X6 below. |
| 12 | Will late Completion disadvantage the *Employer*?<br><br>YES  choose Option X7<br>NO   do nothing | **Discussion**<br>Option X7 allows for liquidated damages to be paid by the *Contractor* if he fails to complete the *works* by the Completion Date. The amount of delay damages should be a currency amount per day, for example, pounds sterling rather than a percentage of the contract value, because the delay damages have to be a genuine pre-estimate of the loss suffered as a result of the late Completion. If delay damages are not chosen, the *Employer* still has the option of suing for damages at large.<br><br>**Action**<br>Choose Option X7 if you want to include for liquidated damages where the *Contractor* achieves Completion later than the Completion Date. | See Contract Data entries for Option X7 below. |
| 13 | Does the *Employer* want to liquidate any damages that may be suffered in consequence of low-performance/substandard work?<br><br>YES  choose Option X17<br>NO   do nothing | **Discussion**<br>Where work is of low standard/substandard, and the standard is clearly stated in the Works Information, the *Employer* may choose to include in the contract liquidated damages that represent the damage he would suffer in consequence of the substandard work. Other options are to insist that the *Contractor* achieves the standard stated in the Works Information, to have someone else correct the Defect and charge the costs to the *Contractor*, or accept the Defect and a quotation from the *Contractor* for reduced Prices.<br><br>This may be particularly relevant on process plant type projects, for example, a gas desulphurisation plant at a power station is required by the Works Information to remove 95% of sulphur from the gases omitted from the flues. The plant only achieves 90%.<br><br>**Action**<br>Choose Option X17 if you wish to liquidate the damages suffered in consequence of the *Contractor* producing low-performance/substandard work. | See Contract Data entries for Option X17 below. |
| 14 | Does the *Employer* want to accept the risk of changes in the law?<br><br>YES  choose Option X2<br>NO   do nothing | **Discussion**<br>Where changes in the law that take place after the Contract Date affect the *Contractor*'s costs, this would generally be the *Contractor*'s risk. For longer contracts where the *Contractor* might not have had warning of changes, the *Employer* may choose to adopt this risk.<br><br>**Action**<br>Choose Option X2 to add a compensation event to the contract for changes to the law if the *Employer* is willing to take the risk of changes to the law that affect the *Contractor*'s costs. | There are no Contract Data entries for Option X2. |

|  | Column A<br>Question | Column B<br>Assistance | Column C<br>Contract Data item |
|---|---|---|---|
| 15 | Does the *Employer* wish to include a partnering option?<br><br>YES  choose Option X12<br>NO   do nothing | **Discussion**<br>The Partnering Option allows for the introduction of managing the contract as if it were a partnership (does not create a legal partnership). The organisations that are a part of the partnership are stated by the *Employer*, and Partnering Information about how the partnership is to be managed is listed in the Works Information. Option X12 also includes its own Contract Data.<br><br>**Action**<br>If the *Employer* wishes to include for the partnering option, Option X12 should be chosen. | See Contract Data entries for X12 below. |
| 16 | Does the *Employer* wish to limit the *Contractor*'s liability under the contract for loss, damage to property or design for a specified period of time?<br><br>YES  choose Option X18<br>NO   do nothing | **Discussion**<br>Option X18 limits the liability of the *Contractor* to the *Employer* for indirect or consequential loss, loss or damage to the *Employer*'s property, latent defects under design and total liability. Although not specifically stated, it may be that these different limitations can be stated individually or included as 'unlimited'.<br><br>**Action**<br>Choose Option X18 to limit the *Contractor*'s liability. | See Contract Data entries for Option X18 below. |
| 17 | Does the *Employer* wish to include Key Performance Indicators?<br><br>YES  choose Option X20<br>NO   do nothing | **Discussion**<br>Option X20 facilitates the inclusion of an Incentive Schedule, which details the targets for Key Performance Indicators. Amounts of money are included in the Incentive Schedule. The *Contractor* is required to report on his performance against each of the Key Performance Indicators to the *Project Manager*.<br><br>**Action**<br>Choose Option X20 to include Key Performance Indicators, targets and monetary incentives for the *Contractor*. | See Contract Data entries for Option X20 below. |
| 18 | Does the HGCR Act 1996 apply to the contract? (UK only)<br><br>YES  choose Option W2<br>NO   choose Option W1 | **Discussion**<br>If the contract is a construction contract in accordance with the UK Act, then adjudication is required to be the first level of dispute resolution and the Act provides the procedures for any such adjudication (generally called 'statutory adjudication'). If the Act does apply to the contract, and Option W2 is not chosen, then Option W1 is not the default position if a dispute requiring adjudication were to arise. The default position is described in the Scheme for Construction Contracts (England and Wales) Regulations 1998 and the Scheme for Construction Contracts (Scotland) Regulations 1998.<br><br>If the HGCR Act 1996 does not apply to the country within which the contract is being executed (the *law of the contract*), or if the contract is not a construction contract as defined by the Act, then Option W1 provides adjudication as a first level of dispute resolution, but the procedures are slightly different from the statutory adjudication described in the Act.<br><br>**Action**<br>Choose Option W2 if the *Employer* thinks that the contract is a construction contract in accordance with the HGCR Act and he does not want the Scheme to apply. Choose Option W1 if the contract is not a construction contract in accordance with the Act.<br><br>Where countries other than the UK have similar Acts or Schemes in place, these could be incorporated through the use of Option Z. | See Contract Data entries for Options W1 and W2 below. |

| | Column A<br>Question | Column B<br>Assistance | Column C<br>Contract Data item |
|---|---|---|---|
| 19 | If the contract is a construction contract in accordance with the HGCR Act 1996, does the *Employer* wish to include the required statements **relating to payment** in the contract?<br><br>YES  choose Option Y(UK)2<br>NO  do nothing (and the Scheme for Construction Contracts will apply) | **Discussion**<br>Option Y(UK)2 relates only to payment. If the contract is a construction contract in accordance with the Act and Option Y(UK)2 is not chosen, then the Scheme for Construction Contracts will be read into the contract. The *Employer* should consider his position in the light of the clauses in both Y(UK)2 and the Scheme and make his decision.<br>**Action**<br>Choose Option Y(UK)2 if the *Employer* thinks that the contract is a construction contract in accordance with the Act and he does not want the Scheme to apply. | The Contract Data entries relating to Option Y(UK)2 are specific to Contract Data section 5 (Payment) and therefore appear in that section below. |
| 20 | Does the *Employer* wish to take into account the Contracts (Rights of Third Parties) Act 1999? (UK only)<br><br>YES  choose Option Y(UK)3<br>NO  do nothing | **Discussion**<br>Option Y(UK)3 covers the introduction of the Contracts (Rights of Third Parties) Act 1999 for use with the ECC when used in England and Wales. This law is covered by common law within Scotland and is therefore not required for contracts for which the *law of the contract* is the law of Scotland. The optional clause should be included in all contracts that came into existence on or after 11 May 2000.<br><br>**Action**<br>If the *Employer* wants to take this Act into account, he should choose Option Y(UK)3. | See Contract Data entry required for Option Y(UK)3. |
| 21 | Does the *Employer* wish to add his own conditions of contract?<br><br>YES  choose Option Z<br>NO  do nothing | **Discussion**<br>Most employers would prefer to include some of their own conditions of contract, such as confidentiality. In general, however, additional conditions should only be used when absolutely necessary to accommodate special needs. The additional conditions of contract should be drafted in the same style as the core and optional clauses, using the same defined terms and other terminology. They should be carefully checked for consistency with the other conditions.<br><br>**Action**<br>Choose Option Z if the *Employer* wants to include for additional conditions of contract that take into account special circumstances for the particular project. | See Contract Data entries for Option Z below. |

## 3.5 Part three of this chapter: choosing optional Contract Data statements

This part of the chapter continues with questions about your chosen contract strategy, guiding you in choices about the optional statements that appear towards the end of the Contract Data.

In part one of the Contract Data, there are a number of 'optional statements' that appear after the Contract Data for the nine sections of core clauses. These optional statements represent aspects of the *Employer*'s contract strategy (the *Contractor*'s in the Engineering and Construction Subcontract (ECS)) and should be decided before issuing tender documentation.

The optional statements are all presented in the Contract Data pro-forma in the ECC as 'if-statements'. If the answer to the if-statement is in the affirmative, then the optional statement is included in the Contract Data. If the answer is in the negative, then the optional statement is not included in the Contract Data.

These optional statements have caused some confusion with *Employers* who are unsure how to present them in their Contact Data. Before detailing the questions and what the optional statements actually mean, below are some guidelines regarding the optional statements and how to present them.

1 If your answer to the 'if' statement presented is 'yes', then:
- delete the 'if' statement
- complete the optional statement
- include the optional statement in the Contract Data in the section under which it should appear in the Contract Data (further guidance is given on this point for the individual statements below).

2 If your answer to the 'if' statement is 'no', then:
- delete the 'if' statement and the Contract Data optional statement.

3 If the 'if' statement appears in both Contract Data part one and Contract Data part two, then:
- remember to choose only one and
- delete the other if-statement and optional statement.

There are only two of these statements, one referring to when the *Employer* wants to receive the first programme and the other referring to who determines the *completion date* – the *Contractor* or the *Employer*.

Any reference to CD1 is a reference to Contract Data part one by the *Employer*. Any reference to CD2 is a reference to Contract Data part two by the *Contractor*.

### 3.5.1 Contract Data by the *Employer*

| | Column A<br>Question | Column B<br>Assistance | Column C<br>Contract Data item |
|---|---|---|---|
| 22 | Does the *Employer* want the *Contractor* to decide the *completion date*?<br><br>YES choose the optional statement in CD2<br>NO choose the optional statement in CD1 | **Discussion**<br>The ECC allows for either the *Contractor* or the *Employer* to decide the *completion date*. In general, where the *Employer* wishes the *Contractor* to tender his programme (see next question) and therefore also tender his realistic *completion date*, the *Employer* would not dictate the *completion date*, but would allow the *Contractor* to dictate it. In most instances, however, the *Employer* has a programme of work and a requirement for the *Contractor* to finish by a specific date.<br><br>**Action**<br>If the *Employer* wants the *Contractor* to decide the *completion date* for the whole of the *works*:<br>- delete the optional statement in Contract Data part one<br>- include the optional statement in Contract Data part two (and leave blank).<br>If the *Employer* wants to dictate the *completion date* for the whole of the *works*:<br>- delete the optional statement in Contract Data part two<br>- include the optional statement in Contract Data part one section 3 and complete the bullet point by adding the date of completion. | **Contract Data part one**<br>- The *completion date* for the whole of the *works* is ...............................<br>**Contract Data part two**<br>- The *completion date* for the whole of the *works* is ............................... |
| 23 | Does the *Employer* want to see a programme with the tender?<br><br>YES choose the optional statement in CD2<br>NO choose the optional statement in CD1 | **Discussion**<br>The *Employer* may receive a first programme from the *Contractor* either at tender stage or after the Contract Date. Where the *Employer* wishes to ensure that the *Contractor* is able to adhere to dictated dates, or where the *Employer* wishes the *Contractor* to dictate his own programme, then the *Employer* would normally choose to see a programme with the tender. Whichever programme is chosen, it will become the Accepted Programme after it has been accepted by the *Project Manager*. There should therefore be very little difference between the 'tender programme' and the 'contract programme'. | **Contract Data part two**<br>- The programme identified in the Contract Data is<br>...............................<br>**Contract Data part one**<br>- The *Contractor* is to submit a first programme for acceptance within<br>........................... weeks of the Contract Date. |

| | Column A<br>Question | Column B<br>Assistance | Column C<br>Contract Data item |
|---|---|---|---|
| | | **Action**<br>If the *Employer* wants a programme with the tender:<br>■ delete the optional statement in Contract Data part one<br>■ include the optional statement in Contract Data part two (and leave blank).<br>If the *Employer* only wants a programme after the Contract Date:<br>■ delete the optional statement in Contract Data part two<br>■ include the optional statement in Contract Data part one section 3 and complete the bullet point by adding the number of weeks after the Contract Date by which time the *Employer* wants to see the programme. | |
| 24 | Is the *Employer* willing to take over the *works* if the *Contractor* completes the *works* before the Completion Date?<br><br>YES  do nothing<br>NO  choose the optional statement in CD1 | **Discussion**<br>Sometimes, if a *Contractor* completes early, the *Employer* might not be willing to take over the *works*, for example, due to constraints on the interaction between the *works* and the rest of the project. If this is the case, then the *Employer* needs to make this clear, since the default position is that the *Employer* takes over the *works* within two weeks after Completion.<br>**Action**<br>If the *Employer* is unwilling to take over the *works* if Completion is before the Completion Date, then include the optional statement in Contract Data part one section 3. | **Contract Data part one**<br>■ The *Employer* is not willing to take over the *works* before the Completion Date. |
| 25 | Does the *Employer* want some work to be completed by *key dates*?<br><br>YES  choose the optional statement in CD1<br>NO  do nothing | **Discussion**<br>The *Employer* may decide to include a date or a series of dates which represent the dates by which the *Contractor* is to meet the condition described in the Contract Data.<br>**Action**<br>If the *Employer* has identified work to be completed by *key dates* then this should be incorporated into the Contract Data. Note that the condition of work described needs to be very clearly defined. | **Contract Data part one**<br>■ The *key dates* and *conditions* to be met are<br><br>*condition* to be met   *key date*<br>1. ....................... .............<br>2. ....................... .............<br>3. ....................... ............. |
| 26a | For contracts where Option Y(UK)2 has not been chosen:<br>The *Employer* is obliged to make payment within three weeks after the assessment. Is this time sufficient?<br><br>YES  do nothing<br>NO  choose the optional statement in CD1 | **Discussion**<br>The trigger for payment is the assessment date, not the payment certificate or the receipt of invoice. Since the payment certificate is given to the *Contractor* one week after the assessment date, and any invoice requested a number of days after the payment certificate, this reduces the time allowed for the accounts department to make a payment. If payment is made four weeks after the assessment day, and assuming the invoice is received on time, then the accounts department has two weeks from receipt of invoice to make a payment.<br>**Action**<br>If the *Employer* wants to increase the period for payment, then include the optional statement in Contract Data part one section 5. | **Contract Data part one**<br>■ The period within which payments are made is<br><br>................................. |
| 26b | For Option Y(UK)2 contracts only:<br>The *Employer* is obliged to make payment within 14 days after the date on which the payment becomes due. Is this time sufficient?<br><br>YES  do nothing<br>NO  choose the optional statement in CD1 (taken from Y(UK)2) | **Discussion**<br>The trigger for payment is the date on which payment becomes due. This date is seven days after the assessment date. The default in Option Y(UK)2 is that the *Employer* has 14 days after the date when payment becomes due to make a payment. This is the same amount of time as the ECC pure gives the *Employer* to pay. If the *Employer* considers this time period to be too short, then he may insert a longer time period in Contract Data part one.<br>**Action**<br>If the *Employer* wants to increase the period for payment, then include the optional statement in Contract Data part one section 5. | **Contract Data part one**<br>■ The period for payment is ................... |

|  | Column A<br>Question | Column B<br>Assistance | Column C<br>Contract Data item |
|---|---|---|---|
| 27 | Are there additional *Employer* risks to be included in the contract?<br><br>YES  choose the optional statement in CD1<br>NO  do nothing | **Discussion**<br>Clause 80.1 of the *conditions of contract* allows additional *Employer* risks to be stated in the Contract Data. The *Employer* should use this when he wishes to limit the *Contractor*'s risk for specific items of risk that the *Employer* is better able to control.<br><br>**Action**<br>If the *Employer* intends to adopt extra risks, choose the optional statement in Contract Data part one and include it in Contract Data part one section 8. | **Contract Data part one**<br>■ These are additional *Employer*'s risks<br>1. .....................................<br>2. .....................................<br>3. ..................................... |
| 28 | Does the *Employer* provide Plant and Materials to the *Contractor*?<br><br>YES  choose the optional statement in CD1<br>NO  do nothing | **Discussion**<br>If the *Employer* free-issues to the *Contractor* certain items of Plant and Materials, the *Employer* might want the *Contractor* to insure for this free-issue material. In such a case, this optional statement would be chosen, with the amount of insurance decided by the *Employer*.<br><br>**Action**<br>If the *Employer* free-issues Plant and Materials to the *Contractor* and the *Employer* wants the *Contractor* to insure for those items, then the optional statement in Contract Data part one should be chosen and included in Contract Data part one section 8. | **Contract Data part one**<br>■ The insurance against loss of or damage to the *works*, Plant and Materials is to include cover for Plant and Materials provided by the *Employer* for an amount of .................... |
| 29 | Is the *Employer* intending to provide any of the insurances in the insurance table?<br><br>YES  choose the optional statement in CD1<br>NO  do nothing | **Discussion**<br>Where the Insurance Table in clause 84.2 of the *conditions of contract* does not represent the way the *Employer* wishes to insure, this optional statement and the next are the mechanism for changing the table. In particular, *Employers* sometimes self-insure, or provide the insurance for the *works* themselves.<br><br>**Action**<br>If the *Employer* provides some of the insurances in the insurance table, choose the optional statement in Contract Data part one and include it in section 8. | **Contract Data part one**<br>■ The *Employer* provides these insurances from the Insurance Table:<br>1. Insurance against......,,,,,<br>   Cover/indemnity is......<br>   The deductibles are.....<br>2. Insurance against.......<br>   Cover/indemnity is......<br>   The deductibles are..... |
| 30 | Is the *Employer* intending to provide additional insurances?<br><br>YES  choose the optional statement in CD1<br>NO  do nothing | **Discussion**<br>Where the Insurance Table in clause 84.2 of the *conditions of contract* does not represent the way the *Employer* wishes to insure, this optional statement and the previous one are the mechanism for changing the table.<br><br>**Action**<br>If the *Employer* intends to provide additional insurances to those already stated in the Insurance Table, then this optional statement should be chosen and included in section 8 of Contract Data part one. | **Contract Data part one**<br>■ The *Employer* provides these additional insurances:<br>1. Insurance against.......<br>   Cover/indemnity is......<br>   The deductibles are.....<br>2. Insurance against.......<br>   Cover/indemnity is......<br>   The deductibles are..... |
| 31 | Does the *Employer* require the *Contractor* to provide additional insurances?<br><br>YES  choose the optional statement in CD1<br>NO  do nothing | **Discussion**<br>The *Employer* could want the *Contractor* to provide additional insurances; for example, if design is a part of the *Contractor*'s scope of work, then the *Employer* is likely to want the *Contractor* to provide professional indemnity insurance.<br><br>**Action**<br>If the *Employer* wants the *Contractor* to provide additional insurances, then the optional statement in Contract Data part one should be chosen and included in section 8 and the required insurance indicated. | **Contract Data part one**<br>■ The *Contractor* provides these additional insurances:<br>1. Insurance against.......<br>   Cover/indemnity is......<br>   The deductibles are.....<br>2. Insurance against.......<br>   Cover/indemnity is......<br>   The deductibles are..... |

|  | Column A<br>Question | Column B<br>Assistance | Column C<br>Contract Data item |
|---|---|---|---|
| 32 | Has the *Employer* chosen arbitration as the second level of dispute resolution?<br><br>YES  choose the optional statement in CD1<br><br>NO  do nothing | **Discussion**<br>The Contract Data requires further information about arbitration to be stated, so if you have chosen arbitration as the *tribunal*, then the arbitration procedure and other elements relating to the arbitration also need to be decided.<br><br>**Action**<br>If the *tribunal* is arbitration, include the optional statement in Contract Data part one section W (whether W1 or W2 is chosen). If the *tribunal* is litigation, there is no need to include the optional statement relating to arbitration. | **Contract Data part one**<br>■ The *arbitration procedure* is ..............<br>■ The place where arbitration is to be held is ...............<br>■ The person or organisation who will choose an arbitrator<br>  ■ if the Parties cannot agree a choice, or<br>  ■ if the *arbitration procedure* does not state who selects an arbitrator, is .......... |

## 3.5.2 Contract Data provided by the *Contractor*

|  | Column A<br>Question | Column B<br>Assistance | Column C<br>Contract Data item |
|---|---|---|---|
| 33 | Is the *Contractor* required to design part of or the whole of the *works*?<br><br>YES  choose the optional statement in CD2<br><br>NO  do nothing | **Discussion**<br>If the *Contractor* is required to design a part of the *works*, then the *Employer* is likely to want to see the *Contractor*'s design proposal at tender stage. The tenderer would therefore be required to include the proposal in his tender.<br><br>**Action**<br>If the *Employer* wants to see the *Contractor*'s design proposal at tender stage, the optional statement in Contract Data part two should be chosen. | **Contract Data part two**<br>■ The Works Information for the *Contractor*'s design is in<br>.....................................<br>..................................... |

## 3.6 Part four of this chapter: all entries in Contract Data part one by the *Employer*

### 3.6.1 Contract Data part one by the *Employer*

This part of the table lists all the information that is needed to complete Contract Data part one by the *Employer*, incorporating the optional statements detailed above.

| | Statement required | Contract Data entry |
|---|---|---|
| **1. General** | | |
| 34 | A list of the main and secondary Options chosen is required (including choices from the amendments Y(UK)2, Y(UK)3 and X12) based on the choices made above. | ■ The *conditions of contract* are the core clauses and the clauses for main Option ................... dispute resolution Option ................... and secondary Options .......... of the NEC3 Engineering and Construction Contract June 2005, with amendments June 2006. |
| 35 | A description of the *works* is required, including all aspects of the work that is to be done, for example, manufacture, supply, delivery to Site, installation, testing and commissioning. | ■ The *works* are the ................... |
| 36 | The name and registered address of the *Employer*. This is the default address for receiving information. | ■ The *Employer* is<br>Name .............................<br>Address ............................... |
| 37 | Name and address of the *Project Manager*.<br>The *Project Manager* should be on Site all the time and is the person by whom most of the decisions are made that impact on the budget or the programme. | ■ The *Project Manager* is<br>Name .............................<br>Address ............................... |
| 38 | Name and address of the *Supervisor*.<br>The *Supervisor* is in charge of testing, quality and title only. | ■ The *Supervisor* is<br>Name .............................<br>Address ............................... |
| 39 | Name and address of the *Adjudicator*.<br>This may be either a person whom you have approached and requested their expertise as adjudicator on potential disputes, or an appointing body who would appoint an adjudicator should a dispute arise. If the latter, then add the words 'to be appointed by' after the Contract Data entry of 'The *Adjudicator* is'.<br>Note that Options W1 and W2 both require the *Adjudicator* to be appointed at the *starting date* by the Parties under the NEC3 Adjudicator's Contract. It would appear, therefore, that a name rather than an adjudicator-nominating body is required to be stated in the Contract Data. There is, however, a default position included, where a Party may ask the *Adjudicator nominating body* included in the Contract Data to choose an adjudicator if one is not identified in the Contract Data. | ■ The *Adjudicator* is<br>Name .............................<br>Address ............................... |
| 40 | The place where the Works Information is to be found should be stated here. The contract places considerable importance on the Works Information. By reference here in the Contract Data, the documents which contain this information are established. By definition, the information includes not only design but also specific requirements regarding how the *Contractor* is to administer and manage the *works*, and therefore several documents may need to be included. An example is:<br>■ The Works Information is in the document entitled 'Works Information' and all documents referred to in it. | ■ The Works Information is in ................................... |
| 41 | The place where the Site Information is to be found should be stated here. An example is<br>■ The Site Information is in the document entitled 'Site Information' and all documents referred to in it. | ■ The Site Information is in ................................... |

| | Statement required | Contract Data entry |
|---|---|---|
| 42 | The information entered here determines the limits of the Site (refer to clause 11.2(15)). This is defined separately from the *working areas* which the *Contractor* identifies in part two of the Contract Data. The definition of the Site is important as it relates to:<br>■ *access dates*<br>■ compensations events, for example, the *Employer* fails to give access of the Site by the date shown on the Accepted Programme<br>■ title to objects/items within the Site<br>■ definition of *Employer*'s risks, i.e. 'use or occupation of the Site'<br>■ Termination – *Employer* may instruct the *Contractor* to leave the Site.<br>A drawing showing the boundaries of the Site is the easiest way to complete this statement. An example is<br>■ The *boundaries of the site* are as shown on drawing DR/xyz/001. | ■ The *boundaries of the site* are<br>................................. |
| 43 | The language refers to the language in which communications are written (see clause 13.1). | ■ The *language of this contract* is ................................. |
| 44 | The law of the contract would be the country in which the Site is situated, by common law; however, larger companies whose headquarters are situated elsewhere might prefer the law to be that of a country other than where the Site is situated. | ■ The *law of the contract* is the law of ................................. |
| 45 | This determines the length of time allowed under the contract for either party to reply to a communication. A communication is defined as any instruction, certificate, submission, proposal, record, acceptance, notification and reply and any other communication which the contract conditions require. It is important to note the content of clause 13.7 which states that any notification required by the contract is communicated separately from other communications. This has particular relevance to the notification of compensation events, which cannot be notified as part of another communication such as the minutes of a meeting. The contract also states specific time periods for the issue of such identified notifications.<br><br>This period of time should be tailored so that both the *Contractor* and the *Project Manager* can respond to a communication within the time period stated. If a longer period of time is required for a specific type of communication, for example, drawings, then two time periods could be stated, for example,<br>■ The *period for reply* is three weeks for drawings and two weeks for all other communications. | ■ The *period for reply* is ................................. weeks. |
| 46 | The definition for Risk Register (clause 11.2(14)) requires a list of risks to be described in the Contract Data. This list may be added to by the *Project Manager* or the *Contractor* through the early warning mechanism. | ■ The following matters will be included in the Risk Register ................................. |

## 2. The *Contractor*'s main responsibilities

| | | |
|---|---|---|
| 47 | There are no Contract Data entries required for this section of the core clauses, but in order to retain the chronological numbering of the Contract Data, you might want to include a statement to this effect. | ■ There are no Contract Data entries required for this section of the core clauses. |

## 3. Time

| | | |
|---|---|---|
| 48 | The *starting date* is one of a series of dates specified by the *Employer* in part one of the Contract Data and around which the *Contractor* plans the *works*. This date is not necessarily the start date on Site, but the date when work for the contract starts, for example, design or fabrication. | ■ The *starting date* is ................................. |
| 49 | Access is when the *Contractor* is given permission to occupy the Site. This might be after the *starting date*, if there is design or manufacture to take place first. If there is only one part of the Site, then this would be described as the 'whole of the Site'. | ■ The *access dates* are<br>Part of the Site    Date<br>1. ....................   ....................<br>2. ....................   .................... |

|  | **Statement required** | **Contract Data entry** |
|---|---|---|
| 50 | The *Contractor* is required to revise his programme every so often (see clause 32.2) and this statement dictates the period of time between revisions. The period stated is typically one month; however, each *Employer* should review the frequency required against the complexity and nature of the *works*. The submission of revised programmes is a demanding activity which the *Contractor* and the *Project Manager* should work on together to achieve. For longer contracts, the interval could be a longer period of time, such as eight weeks. The *Project Manager* always has the option to request a revision outwith the stated period of time. | ■ The *Contractor* submits revised programmes at intervals no longer than ................................. weeks. |
| 51 | This is an optional statement. See your answer to question 22 above as to whether it should be included. | ■ The *completion date* for the whole of the *works* is ................................. |
| 52 | This is an optional statement. See your answer to question 24 above as to whether it should be included. | ■ The *Employer* is not willing to take over the *works* before the Completion Date. |
| 53 | This is an optional statement. See your answer to question 23 above as to whether it should be included. | ■ The *Contractor* is to submit a first programme for acceptance within ................................. weeks of the Contract Date. |
| 54 | This is an optional statement. See your answer to question 25 above as to whether it should be included. Note that the *condition* of work needs to be clearly defined. | ■ The *key dates* and *conditions* to be met are<br><br>*condition* to be met      *key date*<br>1. ...........................   ...............<br>2. ...........................   ...............<br>3. ...........................   ............... |

## 4. Testing and Defects

|  | | |
|---|---|---|
| 55 | The *Contractor* is required to correct Defects free of charge up until the *defects date*, a period of time that might traditionally have been known as the defects liability period or the maintenance period. It would usually be 52 weeks but could be 26 weeks or some other period. On process plant type products it could be 104 weeks. Note that the *defects date* is after Completion of the whole of the *works* and therefore that *sections* of the *works* completed earlier are subject to a longer period of time for the correction of Defects. Note also that the date affects numerous provisions in the conditions as follows.<br><br>■ Defects may not be notified after the *defects date* (clause 42.2).<br>■ Compensation events may not be notified after the *defects date* (clause 61.7).<br>■ The Defects Certificate is issued at the *defects date* or at the end of the last *defect correction period*, whichever is the later (clause 43.3).<br>■ The *Contractor* has an obligation to promptly replace loss of and repair damage to the *works*, Plant and Materials (clause 82.1), and both parties must provide the required insurances until the Defects Certificate has been issued (clause 84.2).<br>■ A payment which is conditional on a *Supervisor*'s test or inspection being successful becomes due at the *defects date* (or end of the last *defect correction period* if later) where the *Supervisor* has not done the test or inspection and the delay to the test or inspection is not the *Contractor*'s fault (clause 40.5).<br>■ X18.3 of Option X18 limits the *Contractor*'s liability to the *Employer* for Defects due to his design after the *defects date* in whatever way is stated by the *Employer* in the Contract Data. | ■ The *defects date* is .......... weeks after Completion of the whole of the *works*. |

| | Statement required | Contract Data entry |
|---|---|---|
| 56 | The *Contractor* is required to correct Defects notified after Completion within a certain period of time after it has been notified to him by the *Supervisor*, or after he has noticed the Defect himself. Although this may be a standard period of time for all Defects, there may be projects where the periods of time would be different for different types of Defects. In this case, an example of the Contract Data statement follows.<br><br>The *defect correction period* is:<br>■ 30 minutes for category A Defects<br>■ 48 hours for category B Defects<br>■ 1 week for category C Defects and<br>■ 4 weeks for category D Defects<br>where Defect categories are described in the Works Information. | ■ The *defect correction period* is ................................. weeks. except that<br>■ The *defect correction period* for ................. is ................. weeks.<br>■ The *defect correction period* for ................. is ................. weeks. |

## 5. Payment

| | | |
|---|---|---|
| 57 | The *currency of the contract* is the currency in which the *Contractor* paid and would generally be 'the pound sterling' for contracts written in the UK. | ■ The *currency of this contract* is the ................................. |
| 58 | The *Project Manager* assesses the amount due at every assessment date. Although the first assessment date takes place to suit the procedures of the Parties, other assessment dates through the contract up to Completion tend to take place every period of time. This period of time may not exceed five weeks and could be:<br>■ 'one calendar month', or<br>■ 'in accordance with the matrix of project dates included in Appendix B to this contract' (where such a schedule includes, for example, assessment dates, certificate dates, invoice dates and payments dates), or<br>■ '4 weeks', in which case the assessment would be slightly earlier every month since some months contain more than four weeks. | ■ The *assessment interval* is ................................. weeks (not more than five). |
| 59 | The *Employer* pays interest if he pays the *Contractor* later than he should. Interest is calculated from the date by which payment should have been made until the date when payment is made and is compounded annually. The fact that the rate is stated as a percentage above the rate of the specified bank allows for the fluctuation of interest rates during the performance of the contract.<br><br>The interest rate is likely to be '2% above the standard base rate of a bank, for example, The Royal Bank of Scotland', or something similar. Try to avoid using averages of many banks if possible, since it increases workload unnecessarily, and introduces conflict, since there are at least three different ways of calculating an average. | ■ The *interest rate* is ............. % per annum (not less than 2) above the ............... rate of the ...................... bank. |
| 60 | This is an optional statement. See your answer to question 26a or 26b above as to whether it should be included. | ■ The period within which payments are made is ........................ (26a).<br>**or**<br>■ The period for payment is ......................... (26b). |

## 6. Compensation events

| | | |
|---|---|---|
| 61 | This statement is used for the compensation event in clause 60.1(13). The comparison that should take place is the weather at this place (on or near the Site) and the weather at the place where it was recorded. This place should therefore be on or as near to the Site as possible in order for the *Contractor* to be able to ascertain the effect of the weather on his work at the Site. | ■ The place where weather is to be recorded is ................................. |

| | Statement required | Contract Data entry |
|---|---|---|
| 62 | These are the aspects of the weather that could affect the *Contractor* in his Providing the Works. They are to be measured at the Site in order to facilitate a comparison with the *weather data* at the place where they were recorded (as stated in item 65 above). An example of additional measurements is wind, where there is cranage; however, if there are no extra measurements, then the last bullet point should be deleted. | ■ The *weather measurements* to be recorded for each calendar month are:<br>■ the cumulative rainfall (mm)<br>■ the number of days with rainfall more than 5 mm<br>■ the number of days with minimum air temperature less than 0 degrees Celsius<br>■ the number of days with snow lying at .............. hours GMT<br>■ and these measurements:<br>.................................<br>................................. |
| 63 | The *Employer* has an opportunity to state who will be taking the weather measurements throughout the period of the contract. This could be the *Contractor*, the *Employer* or a third party. The Met Office has a dedicated section on NEC Contracts. | ■ The *weather measurements* are supplied by<br>................................. |
| 64a | This is the place where the comparative data were recorded. It should also be as near to the Site as possible. Where you have no data, the next Contract Data statement should be used and this one deleted. | ■ The *weather data* are the records of past *weather measurements* for each calendar month which were recorded at<br>................................. and which are available from<br>................................. |
| 64b | Where you have no data to compare the current weather at the Site, you should assume values. In this case, delete the previous Contract Data statement and include this one. | ■ Assumed values for the ten-year return *weather data* for each *weather measurement* for each calendar month are:<br>.............................................<br>.............................................<br>.............................................<br>............................................. |

### 7. Title

| | | |
|---|---|---|
| 65 | There are no Contract Data entries required for this section of the core clauses, but in order to retain the chronological numbering of the Contract Data, you might want to include a statement to this effect. | ■ There are no Contract Data entries required for this section of the core clauses. |

### 8. Risks and insurances

| | | |
|---|---|---|
| 66 | This Contract Data statement correlates with the third statement in the Insurance Table included in clause 84.2 and the amount of insurance required should be stated here.<br><br>ECC3 is specific and refers to the loss or damage having been 'caused by' an activity. | ■ The minimum limit of indemnity for insurance in respect of loss of or damage to property (except the *works*, Plant and Materials and Equipment) and liability for bodily injury to or death of a person (not an employee of the *Contractor*) caused by activity in connection with this contract for any one event is<br>................................. |

|   | Statement required | Contract Data entry |
|---|---|---|
| 67 | This correlates with the fourth statement in the Insurance Table and the amount of insurance required should be stated here.<br><br>ECC3 refers to 'any one event'. | ■ The minimum limit of indemnity for insurance in respect of death of or bodily injury to employees of the *Contractor* arising out of and in the course of their employment in connection with this contract for any one event is<br><br>..................................... |
| 68 | This is an optional statement. See your answer to question 27 above as to whether it should be included. | ■ These are additional *Employer*'s risks:<br>1. ...........................................<br>2. ...........................................<br>3. ........................................... |
| 69 | This is an optional statement. See your answer to question 28 above as to whether it should be included. | ■ The insurance against loss of or damage to the *works*, Plant and Materials is to include cover for Plant and Materials provided by the *Employer* for an amount of<br><br>..................................... |
| 70 | This is an optional statement. See your answer to question 29 above as to whether it should be included. | ■ The *Employer* provides these insurances from the Insurance Table:<br>1. Insurance against.............<br>  Cover/indemnity is............<br>  The deductibles are...........<br>2. Insurance against.............<br>  Cover/indemnity is............<br>  The deductibles are........... |
| 71 | This is an optional statement. See your answer to question 30 above as to whether it should be included. | ■ The *Employer* provides these additional insurances:<br>1. Insurance against.............<br>  Cover/indemnity is............<br>  The deductibles are...........<br>2. Insurance against.............<br>  Cover/indemnity is............<br>  The deductibles are........... |
| 72 | This is an optional statement. See your answer to question 31 above as to whether it should be included. | ■ The *Contractor* provides these additional insurances:<br>1. Insurance against.............<br>  Cover/indemnity is............<br>  The deductibles are...........<br>2. Insurance against.............<br>  Cover/indemnity is............<br>  The deductibles are........... |

## 9. Disputes

|   | Statement required | Contract Data entry |
|---|---|---|
| 73 | There are no Contract Data entries required for this section of the core clauses, but in order to retain the chronological numbering of the Contract Data, you might want to include a statement to this effect. | ■ There are no Contract Data entries required for this section of the core clauses. |
| 73a | No matter whether Option W1 or W2 is chosen, the *Employer* still needs to name the body that will choose an *Adjudicator* should one be needed and the *Employer* needs to decide whether the next form of dispute resolution is arbitration or litigation | ■ The *Adjudicator nominating body* is .........<br>■ The *tribunal* is ......... |

| | | Statement required | Contract Data entry |
|---|---|---|---|
| 73b | | These are optional statements. See your answer to question 32 above as to whether it should be included. If you have chosen arbitration as your *tribunal* then these statements should be included here in the Contract Data. | ■ The *arbitration procedure* is .......... <br> ■ The place where arbitration is to be held is .......... <br> ■ The person or organisation who will choose an arbitrator <br> ■ If the Parties cannot agree a choice or <br> ■ If the arbitration procedure does not state who selects an arbitrator is .......... |

**Statements for Option clauses**

| | | Statement required | Contract Data entry |
|---|---|---|---|
| 74 | **Option A** | There are no Contract Data part one entries required for Option A. | |
| 75 | **Option B** | This is an optional statement. See your answer to questions 1 and 2 above as to whether it should be included. <br> Decide the method of measurement to apply to the *bill of quantities*. | ■ The *method of measurement* is .......... amended as follows .......... |
| 76 | **Option C** | These are optional statements. See your answer to questions 1 and 2 above as to whether they should be included. <br> Decide the share percentages. (See Chapter 2 above for an example.) | ■ The *Contractor's share percentages* and the *share ranges* are <br><br> *share range*      *Contractor's share percentage* <br> less than ..........%    ..............% <br> from.....% to.....%    ..............% <br> from.....% to.....%    ..............,,,% <br> greater than...... %    ..............% |
| 77 | | How often do you want to see forecasts of the out-turn cost? | ■ The *Contractor* prepares forecasts of Defined Cost for the *works* at intervals no longer than .............. weeks. |
| 78 | | Since you will be paying the *Contractor* in other currencies if he has paid a supplier in other currencies, the base exchange rate to be used for conversion into pounds sterling for the purposes of the Fee and the share must be stated. | ■ The *exchange rates* are those published in <br> ................................ on <br> ................................ [date] |
| 79 | **Option D** | These are optional statements. See your answer to questions 1 and 2 above as to whether they should be included. <br> Decide the share percentages. (See Chapter 2 above for an example.) | ■ The *Contractor's share percentages* and the *share ranges* are <br><br> *share range*      *Contractor's share percentage* <br> less than ..........%    ..............% <br> from.....% to.....%    ..............% <br> from.....% to.....%    ..............% <br> greater than...... %    ..............% |
| 80 | | Decide the method of measurement to apply to the *bill of quantities*. | ■ The *method of measurement* is .......... amended as follows .......... |
| 81 | | How often do you want to see forecasts of the out-turn cost? | ■ The *Contractor* prepares forecasts of Defined Cost for the *works* at intervals no longer than ............... weeks. |

| | **Statement required** | **Contract Data entry** |
|---|---|---|
| 82 | Since you will be paying the *Contractor* in other currencies if he has paid a supplier in other currencies, the base exchange rate to be used for conversion into pounds sterling for the purposes of the Fee and the share must be stated. | ■ The *exchange rates* are those published in .......................... on ............................ [date] |
| 83 | **Option E** These are optional statements. See your answer to questions 1 and 2 above as to whether they should be included. Since you will be paying the *Contractor* in other currencies if he has paid a supplier in other currencies, the base exchange rate to be used for conversion into pounds sterling for the purposes of the Fee and the share must be stated. | ■ The *exchange rates* are those published in ................................. on ................................. [date] |
| 84 | How often do you want to see forecasts of the out-turn cost? | ■ The *Contractor* prepares forecasts of Defined Cost for the *works* at intervals no longer than ............... weeks. |
| 85 | **Option F** These are optional statements. See your answer to questions 1 and 2 above as to whether they should be included. | |
| 86 | How often do you want to see forecasts of the out-turn cost? | ■ The *Contractor* prepares forecasts of Defined Cost for the *works* at intervals no longer than ............... weeks. |
| 87 | **Option W** One of Options W1 and W2 is required to be chosen. See your answer to question 18 above as to which one is suitable to your contract. Both Options W1 and W2 allow the facility for someone else to choose an adjudicator, if: ■ the *Adjudicator* is not identified in the Contract Data or ■ the *Adjudicator* resigns or becomes unable to act and ■ the Parties do not wish to or cannot choose an adjudicator jointly. If the Parties are dissatisfied with the *Adjudicator*'s decision, either Party may refer the dispute to a *tribunal*, which would be either the courts or arbitration. This is an optional statement. See your answer to question 32 above as to whether it should be included. This is an optional statement. See your answer to question 32 above as to whether it should be included. This is an optional statement. See your answer to question 32 above as to whether it should be included. | ■ The *Adjudicator nominating body* is ........................... <br> ■ The *tribunal* is ....................... <br> ■ The *arbitration procedure* is ......................... <br> ■ The place where arbitration is to be held is ........................ <br> ■ The person or organisation who will choose an arbitrator ■ if the Parties cannot agree a choice or ■ if the *arbitration procedure* does not state who selects an arbitrator is .................. |

**Statements for Option clauses for ECC3**

| | | |
|---|---|---|
| 88 | **Option X1** This is an optional statement. See your answer to question 9 above as to whether it should be included. Statement for Option X1 only with Options A, B, C and D. | ■ The proportions used to calculate the Price Adjustment Factor are |
| 89 | **Option X2** There are no Contract Data entries required for Option X2. See your answer to question 14 as to whether you have chosen this option. | 0 ...... linked to the index for...... <br> 0 ......                    ...... <br> 0 ...... non-adjustable        ...... <br> ——— <br> 1.00 <br> ■ The *base date* for indices is ................................. <br> ■ The indices are those prepared by ......................... |

| | Statement required | Contract Data entry |
|---|---|---|
| 90 | **Option X3**<br>This is an optional statement. See your answer to question 6 above as to whether it should be included. Statement for Option X3 only. | ■ The *Employer* will pay for the items or activities listed below in the currencies stated. |
| 91 | **Option X4**<br>There are no Contract Data entries required for Option X4. See your answer to question 4 as to whether you have chosen this option. | items and activities .............  other currency .............  total maximum payment in the currency ....................<br>.............  .............  ....................<br>.............  .............  ....................<br><br>■ The *exchange rates* are those published in .......................... on .............................. [date] |
| 92 | **Option X5**<br>This is an optional statement. See your answer to question 7 above as to whether it should be included. Statement for Option X5 only.<br><br>What *sections* of the work are there and when do you want them to be completed? Generally the last section should not be the end of the project but one stage before, since Completion is defined elsewhere. | ■ The *completion date* for each *section* of the *works* is<br><br>section description  *completion date*<br>1.  ...............  ....................<br>2.  ...............  ....................<br>3.  ...............  ....................<br>4.  ...............  .................... |
| 93 | **Where Option X5 and Option X6 are both chosen**<br>This is an optional statement. This Contract Data statement should only be included where Options X5 and X6 are both chosen. If only X5 **or** X6 (or neither) is chosen, then this statement should be deleted.<br><br>The same sections as described in item 120 below should be stated. | ■ The bonus for each *section* of the *works* is<br><br>section description  amount per day<br>1  .................  ...............<br>2.  .................  ...............<br>3.  .................  ...............<br>4.  .................  ...............<br>Remainder of the *works* .............. |
| 94 | **Where Option X5 and Option X7 are both chosen**<br>This is an optional statement. This Contract Data statement should only be included where Options X5 and X7 are both chosen. If only X5 **or** X7 is chosen, then this statement should be deleted.<br><br>The same sections as in item 93 above should be stated. | ■ Delay damages for each *section* of the *works* are<br><br>section description  amount per day<br>1.  .................  ...............<br>2.  .................  ...............<br>3.  .................  ...............<br>4.  .................  ...............<br>Remainder of the *works* .............. |
| 95 | **Option X6**<br>This is an optional statement. See your answer to question 11 above as to whether it should be included. Statement for Option X6 only. | ■ The bonus for the whole of the *works* is ..................... per day |
| 96 | **Option X7**<br>This is an optional statement. See your answer to question 12 above as to whether it should be included. Statement for Option X7 only. Not to be used if Option X5 is also used. | ■ Delay damages for Completion of the whole of the *works* are .................................. per day |

| | **Statement required** | **Contract Data entry** |
|---|---|---|
| 97 | **Option X12**<br>This is an optional statement. See your answer to question 15 above as to whether it should be included. | ■ The *Client* is<br>Name ........................<br>Address ........................<br>........................<br>The *Client's objective* is<br>........................<br>........................<br>........................<br>■ The Partnering Information is in<br>........................<br>........................<br>........................ |
| 98 | **Option X13**<br>This is an optional statement. See your answer to question 3 above as to whether it should be included. Statement for Option X13 only. | ■ The amount of the performance bond is ................ |
| 99 | **Option X14**<br>This is an optional statement. See your answer to question 5 above as to whether it should be included. Statement for Option X14 only. | ■ The amount of the advanced payment is ................ |
| 100 | **Option X15**<br>There are no Contract Data entries required for Option X15. See your answer to question 8 above as to whether you have chosen this option. | ■ The *Contractor* repays the instalments in assessments starting not less than ................ weeks after the Contract Date.<br>■ The instalments are ................ (either an amount or a % of the payment otherwise due)<br>■ An advanced payment bond **is/ is not** required. |
| 101 | **Option X16**<br>This is an optional statement. See your answer to question 10 above as to whether it should be included. Statements for Option X16 only. | ■ The *retention-free amount* is ................<br>■ The *retention percentage* is ................ % |
| 102 | **Option X17**<br>This is an optional statement. See your answer to question 13 above as to whether it should be included. Statement for Option X17. | ■ The amounts for low-performance damages are<br>amount    performance level<br>........... for...........<br>........... for...........<br>........... for........... |
| 103 | **Option X18**<br>This is an optional statement. See your answer to question 16 above as to whether it should be included. Statement for Option X18. | ■ The *Contractor*'s liability to the *Employer* for indirect or consequential loss is limited to ................<br>■ For any one event the *Contractor*'s liability to the *Employer* for loss of or damage to the *Employer*'s property is limited to ................<br>■ The *Contractor*'s liability for Defects due to his design which are not listed on the Defects Certificate is limited to ................ |

| | Statement required | Contract Data entry |
|---|---|---|
| | | ■ The *Contractor*'s total liability to the *Employer* for all matters arising under or in connection with this contract, other than excluded matters, is limited to<br><br>.................................<br><br>■ The *end of liability date* is ................ years after the Completion of the whole of the *works*. |
| 104 | **Option X20**<br>This is an optional statement. See your answer to question 17 as to whether it should be included. Statement for Option X20. This Option may not be used if Option X12 is used. | ■ The *incentive schedule* for Key Performance Indicators is in<br><br>.................................<br><br>■ A report of performance against each Key Performance Indicator is provided at intervals of<br><br>............................. months. |
| 105 | **Option Y(UK)2**<br>This is an optional statement. See your answer to question 19 above as to whether it should be included. The only Contract Data entry required for this Option is included at questions 26a and 26b above. | |
| 106 | **Option Y(UK)3**<br>This is an optional statement. See your answer to question 20 as to whether you have chosen this Option. If there is no third party who may enforce a term, then write 'none'. | ■ term     person or organisation<br>.............   .................................<br>.............   .................................<br>.............   ................................. |
| 107 | **Option Z**<br>This is an optional statement. See your answer to question 21 above as to whether it should be included. | ■ The *additional conditions of contract* are ......................... |

## 3.7 Part five of this chapter: Contract Data part two by the *Contractor*

### 3.7.1 Contract Data part two by the *Contractor*

The *Employer* should remember to choose the correct statements for Contract Data part two, depending on what Options apply to the contract and whether the following apply.

■ Design proposals with the tender.
■ Programme with the tender.
■ The *Contractor* is to dictate the *completion date*.
■ The data for the Schedule of Cost Components might change if you have used Option Z to amend the Schedule of Cost Components in the contract.

The following items relate to each Contract Data part two entry as it appears in the ECC. Some of the entries are decided by the *Employer* and others are completed by the *Contractor* at tender stage.

| | Statement required | Contract Data entry |
|---|---|---|
| 108 | The *Contractor* completes this entry at tender stage.<br>He completes the name of his company and the registered address. | ■ The *Contractor* is<br>Name      ............................<br>Address   ............................<br>................................<br>................................ |
| 109 | The *Contractor* completes this entry at tender stage.<br>The *direct fee percentage* represents the *Contractor*'s profits and overheads that do not already appear in the Schedule of Cost Components for work which is not subcontracted. This entry should be completed no matter what main Option is chosen by the *Employer*. | ■ The *direct fee percentage* is<br>................................ % |

| | Statement required | Contract Data entry |
|---|---|---|
| | The single entry *subcontracted fee percentage* represents the profits and overheads relating to subcontracted work. | ■ The *subcontracted fee percentage* is .................................... % |
| | For Options A and B Defined Cost clause 11.2 (22) is the cost of components in the Shorter Schedule of Cost Components whether work is subcontracted or not. Therefore the *Contractor's* single entry *subcontracted fee percentage* should include and make allowance for The Fees (*direct and subcontracted fee percentages*) tendered by his Subcontractors as well as any profits and overheads he wishes to recover for Subcontracted work. | |
| | For Option C, D and E Defined Cost clause 11.2 (23) is the amount of payments due to Subcontractors for work which is subcontracted. The *Contractor* gets paid what he pays the Subcontractor which includes the Subcontractors Fees (*direct and subcontracted fee percentages*). The *Contractor* then applies his single entry *subcontracted fee percentage* to what he has paid to his Subcontractors. | |
| 110 | The *Contractor* completes this entry at tender stage. | ■ The *working areas* are the Site and ................................. |
| | The *working areas* are very important in defining the *Contractor's* costs and thought should be given as to what should be included in the *working areas*, along with the Site. The *working areas* are those areas within which the *Contractor* can apply for Defined Cost, whether for a compensation event or during the period of the contract. They should therefore not include the *Contractor's* head office, but should include any prefabrication workshops next to the Site. See Chapter 2 of Book 4 on the Schedule of Cost Components for further discussion. | |
| 111 | The *Contractor* completes this entry at tender stage. | ■ The key people are Name............................... Job................................... Responsibilities...................... Qualifications...................... Experience........................... |
| | The people who work on the project can make or break the contract and it is important for the *Employer* to see who he will be working with and their qualifications and experience. If the *Contractor* wishes to replace these people, the substitutes should be as, if not more, qualified and experienced than the key people. The *Contractor* should therefore ensure that the people named at tender stage will be available for the contract. | |
| 112 | The *Contractor* completes this entry at tender stage. | ■ The following matters will be included in the Risk Register: ............................................. ............................................. ............................................. |
| | The Risk Register includes risks which are listed in the Contract Data. The *Employer* has an opportunity to list risks to be included in the Risk Register in his Contract Data part one. The *Contractor* has the same opportunity to list in Contract Data part two what he considers to be risks. | |
| 113 | The *Employer* chooses whether to include this entry. See your answer to question 33 above. | ■ The Works Information for the *Contractor's* design is in ............................................. ............................................. |
| | This is an **optional statement** that should be used where the *Contractor* is providing information for his design, where design is part of his scope of work. The *Employer* should delete the entire bullet point if there is no design involved or if the *Contractor* does not provide information for his design with his tender. | |
| | The *Contractor* completes this Contract Data entry by stating where his design proposal can be located within his tender document; for example, 'The Works Information for the *Contractor's* design is in the document entitled ''*Contractor's* proposal''.' This Works Information will then be included in the contract as Works Information by the *Contractor* and any change to this Works Information made at the *Contractor's* request, to comply with the law or the *Employer's* Works Information, is not a compensation event. | |

| | Statement required | Contract Data entry |
|---|---|---|
| 114 | The *Employer* chooses whether to include this entry. See your answer to question 23 above.<br><br>This is an **optional statement** that is included only when the optional statement for the programme is not included in Contract Data part one. If the optional statement above is chosen and not the one in Contract Data part one, the *Contractor* provides a contract programme with his tender, and does not provide one after the Contract Date. In this case, the bullet point in Contract Data part one should be deleted in its entirety. If the optional statement in Contract Data part one is chosen, the *Contractor* does not provide a programme with his tender, but only provides one after the Contract Date. In the latter case, delete this entire bullet point.<br><br>The *Contractor* should complete this entry by stating where his tender programme can be located in his tender package; for example, 'The programme identified in the Contract Data is in the document entitled "Programme for the *works*".' | ■ The programme identified in the Contract Data is ......................... |
| 115 | The *Employer* chooses whether to include this entry. See your answer to question 22 above.<br><br>This is an **optional statement** that is chosen only if the *Employer* has not chosen a *completion date* in Contract Data part one. If this bullet point is chosen, the *Employer* will have decided that the *Contractor* should decide the *completion date*, and he has also probably asked the *Contractor* for a programme with his tender. In this case, the bullet point in Contract Data part one should be deleted in its entirety. If the *Employer* wants to choose the *completion date*, delete this bullet point in its entirety.<br><br>The *Contractor* should complete this Contract Data entry by stating the date by which he thinks he can achieve Completion. This date should also be included in his programme and should allow for any terminal float; that is, the *Contractor* should include planned Completion as well as Completion in his programme. | The *completion date* for the whole of the *works* is ............................... |
| 116 | **Option A**<br>The *Employer* chooses to include this entry when Option A is the chosen main Option.<br><br>The *Contractor* completes this entry by stating where the *activity schedule* can be found in his tender package; for example, 'The *activity schedule* is in the document entitled "Activity Schedule".'<br><br>The *Contractor* should also include the tendered total of the Prices; that is, the sum of all the lump sum prices in the *activity schedule* in the Contract Data. | ■ The *activity schedule* is<br>.................................<br><br>■ The tendered total of the Prices is ................................. |
| 117 | **Option B**<br>The *Employer* chooses to include this entry when Option B is the chosen main Option.<br><br>The *Contractor* completes this entry by stating where the *bill of quantities* can be found in his tender package; for example, 'The *bill of quantities* is in the document entitled "Bill of Quantities".'<br><br>The *Contractor* should also include the tendered total of the Prices; that is, the sum of all the rates-item product in the *bill of quantities* in the Contract Data. | ■ The *bill of quantities* is<br>.................................<br><br>■ The tendered total of the Prices is ................................. |
| 118 | **Option C**<br>The *Employer* chooses to include this entry when Option C is the chosen main Option.<br><br>The *Contractor* completes this entry by stating where the *activity schedule* can be found in his tender package; for example, 'The *activity schedule* is in the document entitled "Activity Schedule".'<br><br>The *Contractor* should also include the tendered total of the Prices; that is, the sum of all the lump sum prices in the *activity schedule* in the Contract Data. | ■ The *activity schedule* is<br>.................................<br><br>■ The tendered total of the Prices is ................................. |

| | Statement required | Contract Data entry |
|---|---|---|
| 119 | **Option D**<br>The *Employer* chooses to include this entry when Option D is the chosen main Option.<br><br>The *Contractor* completes this entry by stating where the *bill of quantities* can be found in his tender package; for example, 'The *bill of quantities* is in the document entitled ''Bill of Quantities''.'<br><br>The *Contractor* should also include the tendered total of the Prices; that is, the sum of all the rates-item product in the *bill of quantities* in the Contract Data. | ■ The *bill of quantities* is ..................................<br><br>■ The tendered total of the Prices is ................................. |
| 120 | **Option E**<br>There are no Contract Data part two entries required for Option E. | |
| 121 | **Option F**<br>Clause 20.2 states that the *Contractor* will subcontract the *Contractor*'s design, the provision of the Site services and the construction and installation of the *works*, except work which the Contract Data states he will do himself. The *Contractor* should therefore include in his tender the aspects of the *works* which he intends to carry out himself. | ■ Work which the *Contractor* will do himself is .............................<br>activity       *price* (lump sum or unit rates)<br>...................  ...........................<br>...................  ...........................<br>...................  ........................... |

### Data for the full and Shorter Schedule of Cost Components – ECC3

In ECC3, the data is presented and separated according to the main Option used in the contract. For main Options A and B, only the Shorter SCC is used. The first set of data for the SCC to appear in the Contract Data is therefore the data for the Shorter SCC for Options A and B. For main Options C, D and E, the full SCC is used as a default, and the Shorter SCC may be used either by agreement between the *Contractor* and the *Project Manager*, or by the *Project Manager* in making his own assessment. The second, third and fourth sets of data for the SCC to appear in the Contract Data are therefore the data for the full SCC for main Options C, D and E, for both the full and the Shorter SCC for Options C, D and E, and the data for the Shorter SCC for main Options C, D and E respectively.

### Data for the Shorter Schedule of Cost Components

### Main Options A and B only

| | Statement required | Contract Data entry |
|---|---|---|
| 122 | This correlates with clause 41 of the Shorter SCC in ECC3.<br><br>The following payments are separated into four classes.<br><br>42  Payments for cancellation charges arising from a compensation event.<br>43  Payments to public authorities and other properly constituted authorities of charges which they are authorised to make in respect of the *works*.<br>44  Consumables and equipment provided by the *Contractor* for the *Project Manager*'s and *Supervisor*'s office.<br>45  Specialist services.<br><br>In addition, *Contractor*'s accommodation is included in item 2 equipment.<br><br>The people percentage overheads therefore covers item 41, and items 42 to 45 are covered by citing the payment and including documentary evidence. | ■ The percentage for people overheads is ......................... % |
| 123 | This correlates with clause 21 of the Shorter SCC in ECC3.<br><br>The published list should be a list in the public domain rather than an in-house published list. An example is the Civil Engineering Contractors Association (CECA) Dayworks Schedule. | ■ The published list of Equipment is the last edition of the list published by ........................... |
| 124 | This correlates with clause 21 of the Shorter SCC.<br><br>This percentage recognises that the rates in a published list tend to be an all-inclusive rate, thereby including profits and overheads. Because the *Contractor* is able to retrieve his profits and overheads through the *direct fee percentage* or the *subcontracted fee percentage* as applicable and the other overheads stated in the Shorter SCC, the rates in the published list should be reduced. If some rates are reduced by a different percentage to other rates, more than one percentage can be stated, as long as the boundaries of applicability are defined. | ■ The percentage for adjustment for Equipment in the published list is ................................ % (state plus or minus) |

| | Statement required | Contract Data entry |
|---|---|---|
| 125 | This correlates with clause 22 of the Shorter SCC in ECC3.<br><br>The *Contractor* lists here other Equipment that he intends to use, which is not included in the published list. | ■ The rates for other Equipment are<br><br>Equipment    Size/capacity    Rate<br>...............    ...................    ..........<br>...............    ...................    ..........<br>...............    ...................    .......... |
| 126 | These entries correlate with clauses 61, 62 and 63 respectively of the Shorter SCC.<br><br>Note that this data entry refers to costs **outside** the Working Areas (as identified by the *Contractor*). This therefore refers to design in the *Contractor*'s primary place of design, possibly the head office and not any design that takes place on the Site by employees that have been relocated there. Hourly rates could be broken into normal hours (which should be identified) and overtime. Categories of employees could include technician, CAD operator and so on.<br><br>The percentage for overheads should reflect the overheads for the design only; that is, any supervisors, CAD equipment, rent and so on, for that part of the business. This percentage for design overheads should be **exclusive** of profit for that part of the business. This overhead should then **not** be included in the *direct fee percentage* or the *subcontracted fee percentage*, as applicable.<br><br>The categories of employees who are required to travel to and from the Working Areas in the course of their design duties should be stated, so that the *Employer* knows in advance what charges for travel will be made. | ■ The hourly rates for Defined Cost of design outside the Working Areas are<br><br>Category of employee    Hourly rate<br>...............................    .................<br>...............................    .................<br>...............................    .................<br><br>■ The percentage for design overheads is ..................... %<br>■ The categories of design employees whose travelling expenses to and from the Working Areas are included in Defined Cost are ........................<br>................................................... |

## Data for the Schedule of Cost Components

## Main Options C, D and E only

## Data for the full Schedule of Cost Components

## Only used with the full Schedule of Cost Components

The following five bullet points are to be used with the full SCC only for main Options C, D and E. In general, the decision whether to use the full or the Shorter SCC is made by the *Project Manager* and the *Contractor* (in accordance with clause 63.15) for each individual compensation event rather than at the beginning of the contract. The information for both the Shorter and the full SCC should therefore be completed by the *Contractor*.

| | Statement required | Contract Data entry |
|---|---|---|
| 127 | This entry correlates with clause 23 of the full SCC in ECC3.<br><br>The Equipment that is especially purchased for work included in the contract should be listed here. The time-related charge for the Equipment as well as the time period to which the charge relates should also be included. | ■ The listed items of Equipment purchased for work on this contract, with an on-cost charge, are<br><br>Equipment   Time-related   Per time<br>         charge      period<br>...............   ...................   ............<br>...............   ...................   ............<br>...............   ...................   ............ |
| 128 | This correlates with clause 24 of the full SCC in ECC3.<br><br>The *Contractor* lists here other Equipment that he intends to use, together with the rates. | ■ The rates for special Equipment are<br><br>Equipment    Size/capacity    Rate<br>...............    ...................    ............<br>...............    ...................    ............<br>...............    ...................    ............ |
| 129 | This entry correlates with clause 44 of the full SCC in ECC3.<br><br>This percentage represents the *Contractor*'s general costs and preliminaries. Examples of cost elements included in the percentage include messing facilities and site administration but exclude accommodation. | ■ The percentage for Working Areas overheads is .................. % |

| | Statement required | Contract Data entry |
|---|---|---|
| 130 | These entries correlate with clauses 51 and 52 respectively of the full SCC in ECC3.<br><br>Note that this data entry refers to costs **outside** the Working Areas (as identified by the *Contractor*). This therefore refers to manufacture and fabrication in the *Contractor*'s primary place of manufacture and not any temporary fabrication shop set up near the Site (which should be identified as a Working Area). Hourly rates could be broken into normal hours (which should be identified) and overtime. Categories of employees could include welder, labourer and so on. If a supervisor is dedicated to that fabrication shop, then he could be included as a category of employee; otherwise, any supervisors that supervise all jobs and are not dedicated to the particular project should be included as an overhead.<br><br>The percentage for overheads should reflect the overheads in that place of manufacture only; that is, any supervisors, administration, rent and so on, for that building. This percentage for manufacture or fabrication overheads should be **exclusive** of profit for that building. This overhead should then **not** be included in the *direct fee percentage* or the *subcontracted fee percentage* as applicable. | ■ The hourly rates for Defined Cost of manufacture and fabrication outside the Working Areas are<br><br>Category of employee    Hourly rate<br>.............................  ...............<br>.............................  ...............<br>.............................  ...............<br><br>■ The percentage for manufacture and fabrication overheads is<br>...............................  % |

## Data for the Schedule of Cost Components

## Main Options C, D and E only

## Used with both the full and the Shorter Schedule of Cost Components

The following three bullet points are to be used regardless of whether the full or the Shorter SCC is used for main Options C, D and E.

| | | |
|---|---|---|
| 131 | These entries correlate with clauses 61, 62 and 63 respectively of both the full and the Shorter SCC.<br><br>Note that this data entry refers to costs **outside** the Working Areas (as identified by the *Contractor*). This therefore refers to design in the *Contractor*'s primary place of design, possibly the head office and not any design that takes place on the Site by employees that have been relocated there. Hourly rates could be broken into normal hours (which should be identified) and overtime. Categories of employees could include technician, CAD operator and so on.<br><br>The percentage for overheads should reflect the overheads for the design only; that is, any supervisors, CAD equipment, rent and so on, for that part of the business. This percentage for design overheads should be **exclusive** of profit for that part of the business. This overhead should then **not** be included in the *direct fee percentage* or the *subcontracted fee percentage*, as applicable. The categories of employees who are required to travel to and from the Working Areas in the course of their design duties should be stated, so that the *Employer* knows in advance what charges for travel will be made. | ■ The hourly rates for Defined Cost of design outside the Working Areas are<br><br>Category of employee    Hourly rate<br>.............................  ...............<br>.............................  ...............<br>.............................  ...............<br><br>■ The percentage for design overheads is ......................... %<br>■ The categories of design employees whose travelling expenses to and from the Working Areas are included as a cost of design of the *works* and Equipment done outside of the Working Areas are<br><br>...............................................<br>............................................... |

**Data for the Schedule of Cost Components**

**Main Options C, D and E only**

**Data for the Shorter Schedule of Cost Components**

**Only used with the Shorter Schedule of Cost Components**

The following four bullet points are to be used with the Shorter SCC only for main Options C, D and E and only when the Schedule is used by agreement for assessing compensation events or for the *Project Manager* making his own assessment. In general, the decision whether to use the full or the Shorter SCC is made by the *Project Manager* and the *Contractor* (in accordance with clause 63.15) for each individual compensation event rather than at the beginning of the contract. The information for both the Shorter and the full SCC should therefore be completed by the *Contractor*.

The bullet points below generally replace data above that would be used with the full SCC.

| | Statement required | Contract Data entry |
|---|---|---|
| 132 | This correlates with clause 41 of the Shorter SCC in ECC3.<br><br>The following payments are separated into four classes<br><br>42 Payments for cancellation charges arising from a compensation event.<br>43 Payments to public authorities and other properly constituted authorities of charges which they are authorised to make in respect of the *works*.<br>44 Consumables and equipment provided by the *Contractor* for the *Project Manager*'s and *Supervisor*'s offices.<br>45 Specialist services.<br><br>In addition, *Contractor*'s accommodation is included in item 2 equipment.<br><br>The people percentage overheads therefore covers item 41, and items 42 to 45 are covered by citing the payment and including documentary evidence. | ■ The percentage for people overheads is ........................... % |
| 133 | This correlates with clause 21 of the Shorter SCC in ECC3.<br><br>The published list should be a list in the public domain rather than an in-house published list. An example is the Civil Engineering Contractors Association (CECA) Dayworks Schedule. | ■ The published list of Equipment is the last edition of the list published by ............................... |
| 134 | This correlates with clause 21 of the Shorter SCC.<br><br>This percentage recognises that the rates in a published list tend to be an all-inclusive rate, thereby including profits and overheads. Because the *Contractor* is able to retrieve his profits and overheads through the *direct fee percentage* or the *subcontracted fee percentage* as applicable and the other overheads stated in the Shorter SCC, the rates in the published list should be reduced. If some rates are reduced by a different percentage to other rates, more than one percentage can be stated, as long as the boundaries of applicability are defined. | ■ The percentage for adjustment for Equipment in the published list is ................................. % (state plus or minus) |
| 135 | This correlates with clause 22 of the Shorter SCC in ECC3.<br><br>The *Contractor* lists here other Equipment that he intends to use, which is not included in the published list. | ■ The rates for other Equipment are<br><br>Equipment  Size/capacity  Rate<br>...............  ....................  ...........<br>...............  ....................  ...........<br>...............  ....................  ........... |

**Procuring an Engineering and Construction Contract**
ISBN 978-0-7277-5720-3

ICE Publishing: All rights reserved
doi: 10.1680/pecc.57203.115

# Chapter 4
# Works Information guidelines

**Synopsis**

This chapter looks at the Works Information and Site Information, including

- Providing the Works
- what should be included in the Works Information
- separation of the Works and Site Information
- structuring for the Works Information
- interface management
- general rules in drafting the Works Information
- Site Information.

## 4.1 Introduction

The documents within an NEC contract all interact with each other and are therefore required to be drafted in such a manner that they complement each other and are related where indicated. It should also be borne in mind that there is no hierarchy of documents in the ECC. The *conditions of contract* frequently refer to information that is required to be specified within the Works Information and the Contract Data. It is essential that this information appears within these sections of the contract in order to create the links between the documents that make the contract work. If the information does not appear as indicated, ambiguities may arise and the *Employer* or the *Contractor* may be unfairly disadvantaged.

The purpose of this chapter is to identify those clauses within the *conditions of contract* that refer to information that is required to appear in the Works Information. This document also discusses the requirements of these clauses and how they could be incorporated into the Works Information.

The vast majority of disputes arise as a result of the Works Information and the management of it. As the single greatest cause of disputes, it is vital to ensure that the Works Information includes everything the *Employer* requires, and that sufficient time and effort is spent on getting the Works Information as accurate and complete as possible.

This chapter also discusses the Site Information and what is to be included in it. The way in which the Site Information relates to compensation events is also discussed.

## 4.2 Providing the Works

The *Contractor*'s overriding obligation is in clause 20.1, which states that the *Contractor* is required to Provide the Works in accordance with the Works Information. This emphasises the importance of the Works Information: the *Contractor* is only obliged to do those things that the Works Information states he is to do. If it is not in the Works Information, he does not have an obligation to do it.

> The *Contractor*'s overriding obligation is to Provide the Works in accordance with the Works Information.

## 4.3 What should be included in the Works Information

Clauses 11.2(13), 11.2(19) and 11.2(5) all describe what should be included in the Works Information.

1   The *Contractor* is required to complete the *works*, including all incidental works, services and actions required by the contract.
2   The Works Information specifies and describes the *works* or states any constraints on how the *Contractor* Provides the Works. Further notes on the practicalities of drafting the Works Information to achieve these objectives are given in this chapter.
3   A Defect has a very specific meaning under the ECC. A Defect is a part of the *works* that is not in accordance with the Works Information, or a part of the *works* designed by the *Contractor* that is not in accordance with the applicable law or with the *Contractor*'s design that has been accepted by the *Project Manager*. In other words, if the *Employer* requires something to be done, it should be included in the Works Information. If it has not been included, then its absence cannot be a Defect under the contract, and the correction of a 'defect' for its absence would be reimbursable as a compensation event.
4   The description of the *works* is to be found in the Contract Data part one. This description should be as comprehensive as possible, including all aspects of the design, supply, installation, testing and commissioning of the item, particularly bearing in mind the insurance clauses.

Imagine a situation in which the *Supervisor* notifies the *Contractor* that the reinstatement of carriageways on a utility diversion project is not to the carriageway authority's usual standards. However, the Works Information is silent about the reinstatement.

Although it is not to the authority's usual standard, it is **not** a Defect because the test of a Defect is non-conformance with the Works Information. In this situation, if the *works* need to be redone to meet the authority's requirements, the *Contractor* is entitled to a compensation event because the new requirements are a change to the Works Information.

A Defect is a part of the *works* that is not in accordance with the Works Information.

A Defect can be a part of the works designed by the *Contractor* that is not in accordance with the applicable law or with the *Contractor's* design that has been accepted by the *Project Manager*.

## 4.4 Separation of the Works Information and Site Information

To maintain the clarity and simplicity required under the contract it is best to ensure that the Works and Site Information are kept separately from each other.

Traditionally, work specifications and the like have tended to be jumbled together such that the information pertaining to a part of the *works* might be scattered in several locations in the contract documents and could be a mix of both Works Information and Site Information. This could create confusion since the *Contractor's* obligations regarding the Works Information and the Site Information are different.

In an ideal world Works Information and Site Information would be in clearly defined, separate documents. However, we rarely live in an ideal world and in some instances drawings will contain both Works and Site Information. In such instances this needs to be clearly identified and some distinction shown for these drawings, perhaps by means of an asterisk or other label to show that these drawings contain a mix of both Works and Site Information.

They will then need to be listed in both the Works and Site Information drawing list, marked with the special reference chosen, for example:

**Works Information – Example Drawing List**

| Drawing No. | Description | Rev |
|---|---|---|
| 0100 | New Work | 0 |
| 0200* | Work to Existing Factory | 0 |

Note: *Denotes a drawing containing both Works and Site Information

**Site Information – Example Drawing List**

| Drawing No. | Description | Rev |
|---|---|---|
| 500 | Plan of Existing Factory | A |
| 0200* | Work to Existing Factory | 0 |

Note: *Denotes a drawing containing both Works and Site Information

The Works Information and Site Information should be kept in separate documents.

## 4.5 Where the Works Information fits into the contract documents

The typical order of NEC3 ECC contract documents is

- Form of Contract/Articles of agreement
- Contract Data parts one and two
- Pricing Document (e.g. *activity schedule* or *bill of quantities*)
- Works Information
- Site Information.

Note that the Works Information is separate from both the information about the Site and the contract particulars as stated in the Contract Data.

---

The *Employer*'s Works Information and Site Information could comprise the following:

| | |
|---|---|
| **Volume 1A** | Works Information<br>Section A General<br>(May include typical traditional preliminary items) |
| **Volume 1B** | Works Information<br>Specification Sections B to Y<br>(Traditional materials and workmanship specifications) |
| **Volume 1C** | Works Information<br>Specification Sections Z<br>*Employer*'s requirements for<br>*Contractor*'s designed work<br>(The ECC relies on the Works Information to clearly demarcate *Employer*- and *Contractor*-designed *works*) |
| **Volume 1D** | Works Information<br>Safety, health and welfare requirements |
| **Volume 1E** | Works Information<br>Reinforcement bending schedules |
| **Volume 2** | Site Information |
| **Drawings** | Refer Appendix A of specification Section A and Appendix A of Site Information |

---

If the *Contractor* is to provide Works Information for his design, then there will be an optional statement in Contract Data part two, which requires him to identify where the Works Information for his design can be found, for example:

**Document Reference Contract 123/2003**

Works Information for the *Contractor*'s design

This information will be sent with his tender and will form part of the Works Information for the contract once negotiations have been concluded.

## 4.6 Structuring the Works Information

The Works Information specifies and describes the *works*. Therefore, it will cover items such as

- General Works Information (Preliminaries)
- Work Specifications
- *Employer*'s requirements for *Contractor*-designed work
- safety, health and welfare requirements
- reinforcement schedules
- drawings.

It will state the restraints, restrictions and obligations of the Parties. The Works Information may also include Works Information by the *Contractor*, where the contract is a design-and-build contract and the *Contractor* has tendered a *Contractor*'s proposal that meets the *Employer*'s requirements stated elsewhere in the Works Information. The *Contractor*'s proposal forms part of the Works Information because it details how the *Contractor* intends to Provide the Works, but changes to it are not compensation events (clause 60.1(1)). Works Information by the *Contractor* should be kept separate from Works Information by the *Employer*.

Further details of some of the items that could be included in the Works Information follow.

### 4.6.1 Works Information – general

A typical layout and type of items for this section is as follows.

- Brief description of the *works*.
- Provision, content and use of documents.
- Management of the *works*.
- Management and staff.
- Quality Assurance.
- Security and protection.
- Facilities and services to be provided by the *Contractor* for use by the *Employer* and Others.
- Facilities and services to be provided by the *Contractor* for his own use.
- Facilities and services to be provided by the *Employer* for use by the *Contractor*.
- Work by Others with whom the *Contractor* shares the Working Areas.
- Property in excavated materials or demolished buildings.
- Marking Equipment, Plant and Materials.

Note that this section does not include contract particulars, as it would do with a traditional contract, since the contract information is included in the Contract Data. The contract information given in the Contract Data is not Works Information or Site Information and there should be no contractual information or commercial terms in the Works Information.

On repetitive-type work or on framework-type arrangements where a number of similar-type contracts are to be let, it may be more appropriate to separate the Works Information into

1  generic Works Information (common to all contracts) and
2  contract-specific Works Information.

This type of approach can save much time and effort and duplication of information.

### 4.6.1.1 Description of the *works*

A general description for the whole of the *works* should be given, including the *Employer*'s objectives, followed by a detailed description of the scope of the *works*. The failure to detail here a part of the *works*, even if it is given in great detail elsewhere in the document, could be construed as being an ambiguity or inconsistency, which would give rise to a compensation event.

The *works* could also include *Contractor*-designed elements of the *works*, which should also be referred to, although they are detailed later in the Works Information.

The general description of the *works* should also state the outline particulars of the design for those parts of the *works* not fully designed prior to award of the contract and prior to any assumptions the *Contractor* should have allowed for in his tendered price.

This section should also include

- general arrangement and location drawings

■ references to working/production and other detailed drawings, specification, models, and other means to describe the parts of the *works* designed by the *Employer*

■ a statement of any constraints on how the *Contractor* Provides the Works, for example, restrictions on access, sequences of construction.

**4.6.1.2 Management of the works**

This will cover aspects such as site progress meetings where the following issues may be discussed.

■ Administration of insurances.
■ Industrial relations.
■ Risk management.
■ Accounts and records to be kept (depending on the main Option chosen).
■ Procurement procedures (depending on the main Option chosen).

It should state all the specific requirements of the *Employer*'s team, including IT compatibility and the like.

**4.6.1.3 Programme**

Clause 31.2 of the *conditions of contract* includes for any other information stated in the Works Information that the *Contractor* is required to show in his programme. On multi-contract projects, this could include boundary data, foundation design data and similar information relating to his design for the use of the *Employer* and other contractors. The Works Information should clearly state that these dates be shown on the programme. Information about the management of the programme could also be included here.

**4.6.1.4 Completion**

Clause 11.2(2) requires the level of completion to be specified by the *Employer*. There is no 'substantial completion', 'partial completion' or 'mechanical completion' as in traditional forms of contract. This means that Completion must be clearly defined in the Works Information so that the *Contractor* knows when Completion is reached and so that the *Project Manager* knows when to certify Completion.

ECC3 clause 11.2(2) includes a default position whereby if the work which the *Contractor* is to do by the Completion Date is not stated in the Works Information, Completion is when the *Contractor* has done all the work necessary for the *Employer* to use the *works* and for Others to do their work. This lessens the onus on the *Employer* to ensure that a description of Completion is included in the Works Information.

Completion is a status that occurs when the *Project Manager* decides that it has occurred (clause 30.2). In order to decide that Completion has occurred, the *Project Manager* would need a definitive statement indicating what has to be achieved before Completion has occurred. It is worth noting that there is no obligation as with other contracts for the *Contractor* to notify Completion.

---

Completion is when the following has been done by the Completion Date:

(*a*) list of the work required to be done by the Completion Date for the whole of the *works* and for each of the *sections*;

or

Completion is when the *Contractor* has done everything required to Provide the Works except:

(*a*) list of work, which can remain undone at the Completion Date.

---

Completion cannot occur if Defects exist that would prevent the *Employer* from using the *works* (clause 11.2(2)) or Others from doing their work (clause 11.2(2)). In addition, it is likely that there will be Defects that have not yet been corrected by Completion because the *Contractor* may use discretion as to which notified Defects to correct before Completion (clause 43.2). This part of the definition becomes superfluous if clause 43.2 is amended as recommended (see Chapter 3 of Book 3).

**4.6.1.5 Take over**

**Before the Completion Date**
The *Employer* can choose not to take over the *works* if Completion occurs before the Completion Date. If this is the case, the *Employer* states this in Contract Data part one.

**Use of part of the *works***
According to clause 35.2, the *Employer* is deemed to have taken over any part of the *works* where he uses that part before Completion unless the reason for the use is stated in the Works Information. If it is envisaged that the *Employer* might use a part of the *works* without wishing to have taken it over, then the reason for the use should be stated in the Works Information. When the *Employer* takes over the *works* early, delay damages should be reduced.

---

The *Employer* will require the use of the following part of the *works* prior to Completion:

1   the new car park areas as shown on drawing ABC/2002/125A,
2   the new entrance to the Site as shown on drawing ABC/2002/135B

for the following reason .....................................................................

---

**4.6.1.6 Accounts and records**

The *Contractor* is required to keep other records as stated in the Works Information (clause 52.2 for Options C, D, E and F. The Works Information should therefore state what records are to be kept, other than those in clause 52.2.

---

The following is a list of accounts and records to be kept by the *Contractor*:

(*a*)   labour records,
(*b*)   plant records,
(*c*)   all invoices received from Subcontractors.

---

**4.6.1.7 Procurement procedures**

The *Project Manager* is obliged to disallow costs that were incurred because the *Contractor* did not follow an acceptance or procurement procedure stated in the Works Information (clause 11.2(25) for Options C, D and E; clause 11.2(26) for Option F). The Works Information should therefore state those procedures, where required.

---

The acceptance or procurement procedures followed by the *Contractor* are as follows:

(*a*)   .....................................................................
(*b*)   .....................................................................

---

**4.6.1.8 Bonds**

**Performance bond**
If Option X13 has been chosen, the form of the performance bond should be incorporated into the Works Information.

**Parent company guarantee**
If Option X4 has been chosen, the form of the parent company guarantee should be incorporated into the Works Information.

**Advanced payment bond**
If Option X14 has been chosen and a bond is required, the form of the advanced payment bond should be incorporated into the Works Information.

**4.6.1.9 Subcontractors**

It is necessary to state whether parts of the *works* may or may not be subcontracted. There is no nominated subcontracting in the NEC. If the *Employer* requires a part of the *works* to be performed by a specific contractor, it is preferable to contract directly with that *Contractor* where the *Employer* can maintain control, and where the *Employer* should not then be subject to compensation events as a result of enforcing a subcontractor's use. The *Employer* will however take on the interface risk between the *Contractor* and the contractor directly employed by the *Employer*. Where parts of the *works* may be subcontracted, the Works Information should state what Subcontractors are acceptable to the *Employer*.

---

The *Contractor* may not subcontract the *works*,

or

The following work may not be subcontracted:

(*a*) ........................................................................................................
(*b*) ........................................................................................................

or

The *Contractor* may subcontract part of the *works*,

or

The *Contractor* subcontracts the following work:

(*a*) ........................................................................................................
(*b*) ........................................................................................................

The following Subcontractors are acceptable to the *Employer* for performing the work required by this contract:

(*a*)  Kitchen Installations:
    1  Kitchen World Limited
       1 Cupboard Way, Croydon
    2  Mersea Limited,
       1 Sea Road, Ipswich
(*b*)  Furniture and Fittings.

---

The ECC has no nomination process as in traditional contracts.

The *Employer* has two alternatives:

1  Direct appointment with other contractors.
2  Include a list of approved suppliers/contractors in the Works Information.

**4.6.1.10 Quality Assurances**

Quality Assurances will cover

- samples of Plant, Materials and workmanship
- acceptance of Plant and Materials
- compliance with recognised good practice
- compliance with manufacturer's recommendation
- ordering and supply of Plant and Materials
- handling, storing and fixing
- storage of Plant and Materials
- setting out the *works* (this is not covered within the contract conditions as it is with traditional contacts, so this must be stated in the Works Information)

■ statements of how the *Contractor* plans to do the work – detailed requirements

■ instrumentation.

**4.6.1.11 Security and protection**

Detail the *Employer*'s security arrangement or requirements and details of protecting existing services, structures, etc.

Control of noise, cleanliness of access roads.

**4.6.1.12 Facilities and services to be provided by the *Contractor* for use by the *Employer* and Others**

According to clause 25.2, the *Employer* and the *Contractor* provide services and other things as stated in the Works Information. The Works Information should therefore state the interfaces between contractors on site. Any services provided by one contractor to another or by or to the *Employer* should also be stated as well as responsibility for the provision and maintenance of facilities such as

■ access roads, scaffolding, cranes and hoists, water

■ accommodation for the *Employer* and *Project Manager*'s staff

■ facilities and services for 'common use'

■ fences, screens and hoardings

■ temporary facilities.

> The following services are provided on Site by the *Employer* during the periods stated. [storage, power supplies, water, compressed air, telephone]
>
> (*a*) ..............................................................................................
> (*b*) ..............................................................................................

**4.6.1.13 Facilities and services to be provided by the *Contractor* for his own use**

This includes everything not provided by the *Employer*, such as

■ offices and storage sheds

■ welfare facilities and services

■ measures for tidiness at work sites

■ name board and advertising

■ power and lighting

■ water

■ drainage

■ communications

■ temporary lighting

■ temporary facilities and services coordination

■ temporary roads, hardstandings and crossings

■ fences, screens and hoarding.

It should be noted this list is not exhaustive and that a list of this sort is not recommended for inclusion in the Works Information. See below for a simple way to include all *Contractor*-provided facilities.

> The following services are provided by the *Contractor* during the periods stated.
>
> (*a*) ..............................................................................................
> (*b*) ..............................................................................................

**4.6.1.14 Facilities and services to be provided by the *Employer* for use by the *Contractor***

This is where the *Employer* states the particulars of the facilities and services that will be provided by the *Employer* at each site. This will cover items such as connection points or locations where water and electricity supplies can be utilised.

**Free-issue items**
Some *Employers* free-issue Plant and Materials to the *Contractor* in order to reduce the cost of the project, where the *Employer* has agreements with suppliers who provide items to the

*Employer* at lower than trade prices. If this is the case, the Works Information should be very clear about what it is that the *Employer* is going to provide in order to avoid the situation where each Party thought the other Party was going to supply the item. Providing Plant and Materials may affect insurances. Another issue that needs to be considered is who is responsible for delivery/collection.

> This section describes what the *Employer* is to supply specifically for the purpose of the *works*. The *Contractor* is to supply everything else required to Provide the Works.

**4.6.1.15 Work provided**

According to clause 31.2, the *Contractor* shows on each programme, among other things, the order and timing of the work of the *Employer* and Others either as stated in the Works Information or as last agreed, and any other information the Works Information requires the *Contractor* to show.

The Works Information should therefore state what work the *Employer* would do or Others not directly involved in this contract. Also included should be those things other than those stated in clause 31.2 that are required to be shown on each programme.

> The work of the *Employer* and Others to be included in the programme are:
>
> (*a*) ...........................................................................................................
> (*b*) ...........................................................................................................
>
> The *Contractor* shows on his programme submitted for acceptance:
>
> (*a*) ...........................................................................................................
> (*b*) ...........................................................................................................

It should be noted that if the *Employer* or Others do not work within the conditions stated in the Works Information, the *Contractor* is entitled to a compensation event in accordance with clause 60.1(5).

**4.6.1.16 Working with Others**

This section details operations or other *works* within the Working Areas by Others under separate arrangements with the *Employer*. A good way to deal with this is to list each operation or work by Others in this section, giving a brief description of each, where when and what they will be doing.

**Cooperation clause 25.1**
If the *Contractor* is required to cooperate with Others then the Works Information should state what information he needs to obtain and provide for them in connection with the *works*.

It is also important to ensure that it is clearly stated how the *Contractor* will share the Working Areas with Others.

**Approval from Others – clause 27.1**
This clause states that the *Contractor* obtains approval of his design from Others where necessary.

This clause therefore requires that the Works Information is clear on whom the *Contractor* needs to obtain approval from for his design.

**Sharing the Working Areas**
According to clause 25.1, the *Contractor* shares the Working Areas with Others as stated in the Works Information. The Working Areas are the Site and other areas defined by the

*Contractor*, and Others are people such as other contractors and other people not directly connected with this contract. The Works Information should therefore include a statement about who else would be working on the Site with whom the *Contractor* might come into contact.

---

The *Contractor* shares the Working Areas with Others as follows:

(*a*)  Which person/body?
(*b*)  For what period of time?
(*c*)  For what part of the Working Areas?

---

It may also be useful to show this information in the form of a restraints programme, which indicates the key milestone dates and indicates other constraints such as works by other contractors.

**4.6.1.17 Title to materials from excavation and demolition**

The *Employer* should state here the title to materials arising from excavations and demolitions.

According to clause 73.2, the *Contractor* has title to materials from excavation and demolition only as stated in the Works Information. The Works Information should therefore state that either the *Contractor* has no title or all title, or else a statement of any materials from excavation and demolition to which the *Contractor* will have title.

---

The *Contractor* has no title to materials from excavation and demolition,

or

The *Contractor* has title to all materials from excavation or demolition,

or

The *Contractor* has title to materials from excavation and demolition as follows:

(*a*)  ...........................................................................................................
(*b*)  ...........................................................................................................

---

Consideration also needs to be given in the Works Information to

- storage arrangements
- salvage of reclaimed materials
- credit to the *Employer* for the sale of such materials.

---

**Salvaged/reclaimed materials**
The *Contractor* shall carefully take down and take to the *Employer*'s store the existing oak panelling in the old boardroom.

The *Contractor* shall provide a method statement detailing how he proposes to dismantle the panelling, together with a sequence of operations for the reassembly of the panelling. This will include the identification and marking of the panelling, the measures taken to protect the panelling during the dismantling process and the method of protection for storage.

> Arrangements have been made for the panelling to be stored in the *Employer*'s stores at:
>
> Building 305
> Field Road
> Newbridge
> NW19 5UR
>
> Contact the Stores Manager on: 01488 555820 extn 1234

**4.6.1.18 Materials off Site**

According to clause 41.1, the *Contractor* does not bring to the Working Areas those Plant and Materials that the Works Information states are to be tested or inspected before delivery. The Works Information should therefore state which Plant and Materials are to be tested or inspected before delivery to the Working Areas.

It should also be noted that there is no payment for Plant and Materials on Site unless the *Employer* has stated that he is willing to pay for them.

According to clause 71.1, the *Contractor* prepares for marking Equipment, Plant and Materials that are outside the Working Areas, as the Works Information requires. The Works Information should therefore state which items are to be prepared for marking, and how this is to be done.

> The *Contractor* prepares for marking Equipment, Plant and Materials that are outside the Working Areas as follows:
>
> (*a*) ................................................................................................
> (*b*) ................................................................................................

**4.6.2 Work Specifications**

Work specifications should detail exactly what it is that the *Contractor* is to provide. It is therefore very important that the specifications are clear, concise and are unambiguous about what it is that the *Contractor* is to provide and price for.

It could include, for example, specifications for

- site clearance
- earthworks
- brickwork
- concrete
- reinforcement
- drawings.

For *Employer*-designed work and performance, it will include specifications for *Contractor*-designed work.

Where Plant, Materials and workmanship specifications are dictated by the *Employer*, these should be included in the Works Information, as well as requirements for delivery and storage before their incorporation in the *works* and the provision of spares, operating manuals and the like.

**4.6.2.1 Specific requirements for the methods, sequence and timing of the *works* and use of the Site**

This section of the Works Information will detail everything that will affect how, where, when and what will be done.

Great care needs to be taken when writing this section to ensure that it does not conflict, creating ambiguities or inconsistencies between this section and the remainder of the Works Information.

This section will cover such information as

- design constraints on the methods, sequence and timing of the *works*
- timing of the *works*
- restrictions on the use of the work site
- other restrictions and requirements
- working hours
- night working restrictions
- noise levels
- *Employer*'s approval of publications
- excavation of sand and gravel ⎫
- explosives. ⎭ put in specifications

**4.6.2.2 Tests and inspections**

The *Supervisor* is responsible for the quality of the *works* and for ensuring that the *works* comply with the Works Information.

He is responsible for testing and inspection of the *works* as the Works Information requires. It is therefore extremely important that the Works Information spells out clearly all tests and inspections, which must have been carried out before completion of an item, or section of the *works* will be given. The Works Information should also state who provides the facilities and materials for the testing.

According to clause 40.1, clause 40 only tests and inspections required by the Works Information and the applicable law. It is therefore important to ensure that all the tests and inspections that are applicable to the *works* are stated in the Works Information. Where any general specifications quoted detail tests required, and those are considered to be sufficient, there is no need to repeat them. It is necessary, however, to ensure that those tests comply with the requirements in the core clauses.

---

(*a*)  Description of test:
- objective, procedure and standards used in the test
- performed at what stage of the work
- method used
- materials, facilities and samples required to be provided by the *Contractor* and the *Employer* for tests and inspections
- time required to perform the test
- person to conduct the test
- acceptable results and deviations
- any documents provided for and produced as a result of the test.

---

According to clause 40.2, the *Contractor* and the *Employer* provide materials, facilities and samples for tests and inspections as stated in the Works Information. Ensure that the tests stated in any general specifications include this information. If not, include it in the Works Information, with reference to the relevant test and specification. For tests that are described in the Works Information, refer to the list above for an indication of what should be detailed.

If Option X17 is chosen, the Works Information should include details of tests to measure the performance of the *works* in operation for which low-performance damages are specified.

---

The following tests are done to measure the performance of the *works*:

(*a*)  ........................................................................................................

(*b*)  ........................................................................................................

---

**4.6.3 Design**

Clause 21.1 states that if the *Contractor* is required to design elements of the *works*, the Works Information should state what parts of the *works* the *Contractor* is to design, and what the interfaces with the *Employer* are.

For contracts with little *Contractor*'s design, a list of what is left to be designed by the *Contractor* should be provided. For more comprehensive design-and-build contracts, a list of what has been designed by the *Employer* should be given, with the *Contractor* being made responsible for designing the remainder.

---

**Contractor's design**
The *Contractor* designs the following parts of the *works*:

(a) ................................................................................................
(b) ................................................................................................

or

The *Employer* designs the following parts of the *works*. The *Contractor* designs everything else:

(a) ................................................................................................
(b) ................................................................................................

---

Also state the interfaces with those parts of the *works* designed by the *Employer*, and the design criteria. A design brief or performance specification for those parts of the *works* to be designed by the *Contractor* should be included in the Works Information. The design brief for the parts of the *works* to be designed by the *Contractor* should cover the following and any other relevant matters.

- Size or space limitations.
- Design standards and codes of practice (on what size paper and type of paper, how many copies to be made, to be retained by whom, etc.).
- Plant, materials and workmanship specifications including references to relevant standards.
- Loading and capacity requirements.
- Operational performance requirements and design life.

According to clause 21.2, the *Contractor* submits the particulars of his design, as the Works Information requires, to the *Project Manager* for acceptance. These elements of how and when particulars are submitted should be included in the Works Information, including requirements for certification and/or checking.

---

**Example**
The following are procedures that the *Contractor* is to follow in carrying out his design:

(a) ................................................................................................
(b) ................................................................................................

---

According to clause 22.1, the *Employer* may use and copy the *Contractor*'s design for any purpose connected with construction, use, alteration or demolition of the *works* unless otherwise stated in the Works Information and for other purposes as stated in the Works Information.

The Works Information should therefore state any additional purposes for which the *Employer* might wish to put the *Contractor*'s design to use, for example, using it to build the same thing again elsewhere.

> The *Employer* may use and copy the *Contractor*'s design for the following purposes:
>
> (*a*) ...............................................................................................
> (*b*) ...............................................................................................

It should also cover design sanctioning procedures, and cover details of the design particulars that the *Project Manager* wishes to see together with the submission details and certification/acceptance procedure, and deal with *Contractor*'s changes.

**4.6.3.1 Design of Equipment**

According to clause 23.1, the *Contractor* may be required to submit particulars of the design of an item of Equipment to the *Project Manager* for acceptance. If the *Employer* requires the *Contractor* to design items of Equipment, the details of the Equipment and the design particulars should be detailed in the Works Information in the same way as for the design of the *works*.

**4.6.4 Health and safety requirements**

Clause 27.4 requires health and safety requirements to be stated in the Works Information. Since the *Contractor* has to adhere to the law (as does the *Employer*), it is not necessary to reiterate parts of any statutes in the Works Information. This clause therefore applies only to those requirements that are **additional** to those required by the law. The *Employer* may have specific in-house requirements that require a greater level of health and safety than that imposed by the law. For example, *Employer*-specific requirements for airports, rail, petrochemical, power station sites. In this case, those requirements would be stated in the Works Information or incorporated by reference, always ensuring that inconsistency between documents or between the law and the *Employer*'s requirements are eradicated.

A health and safety plan as required by the law (CDM Regulations) should be included.

**4.6.5 Reinforcement schedules**

These will appear as appendices to the Works Information.

**4.6.6 Drawings**
**4.6.6.1 Introduction**

The designer needs to ask himself some basic questions about the drawings.

Do the drawings give the *Contractor* sufficient information to Provide the Works? For example

- Are all associated relevant drawings referred to?
- Are the appropriate parts of the Works Information referred to?
- Is the terminology correct, for example, *Project Manager*, *Supervisor*?
- Are there any woolly words, for example, 'to the satisfaction of the *Project Manager*'?
- Are the drawings cross-referenced to any boundary limits given?

**4.6.6.2 The importance of labelling**

With any contract, care should be taken with labelling. If we look at the example in Figure 4.1, it shows a gas main, which runs between two points A and B. Unfortunately the labelling indicates it as an existing and a 'new' 180 mm pipe. In this simple example, it is evident that the intent is for a new 180 mm polyethylene gas pipe between points A

**Figure 4.1** New gas main

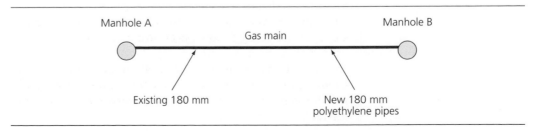

129

and B. However, the draughtsman's carelessness in labelling gives rise to an inconsistency under clause 17 and the *Project Manager* has to issue an instruction to clarify the Works Information, which is a compensation event.

Even before the *Contractor* commences work on site he could have many compensation events, which will give rise to price and time changes if care is not taken in the documents.

## 4.7 Interface management

On multi-contract projects one of the biggest challenges (or risks depending on your perspective) is the management of the interface between the various contractors. Like most risks, this one is best avoided if at all possible, but the reality is that on most major projects with demanding programmes some interface between contractors is inevitable. The ECC makes it clear who carries the risk in the event of an error in the management of the interface between the *Contractor* and other contractors. Clause 60.1(5) makes it a compensation event, and therefore an *Employer*'s risk, if 'other contractors do not work within the times shown on the Accepted Programme or do not work within the conditions stated in the Works Information or 'carry out work on the Site that is not stated in the Works Information'.

The ECC3 Guidance Notes offer some good advice for situations where the interface between the *Contractor* and Other contractors (or other bodies) is complex. It states:

> 'It is important that work [involving interfaces] is planned and programmed as far as possible **before** the start of the contract. The start dates for work should be agreed together with details of the work, its likely duration and facilities required to be provided by and for the *Contractor*. Details of the obligations of the parties at each interface and the timing and programming arrangements should be agreed. This information may conveniently be provided in the form of interface schedules in the Works Information to ensure that arrangements in the different contracts are "back to back". It should also be stated which party is to supply and maintain access (e.g. scaffolding, lifting equipment for plant), resources (for testing etc.) and other services (such as power and water supply).' (ECC Guidance Notes page 42 second paragraph of clause 25.2)

On multi-contractor projects several contractors may be sharing the same work site or one contractor may be following another, in which case the first contractor might set up and leave the site set up on completion of his work. Difficulties nearly always occur with such interfaces. It should be clear in the Works Information what is to be left at take over and how it is to be left.

It is important that not only the permanent works have a clear definition of when an item of work will be complete, but also any temporary, or temporary permanent works, for example, site accommodation, temporary lighting, electrics, ventilation, water supplies, access scaffolding, safety handrails, etc.

It is important that you match what is left by one *Contractor*, to what the next *Contractor* expects to find.

For instance, the Works Information states:

**Shaft 'A' – 10.4 m diameter**
On Completion of the *works* the *Contractor* shall leave an access scaffold in the shaft for use of the follow-on *Contractor*.

You may think this is simple enough, but consider the following scenario. This *Contractor* will leave an access scaffold, but what are our requirements? What standards must it meet? How will we prove to the next *Contractor* that he can take it over?

We can see immediately that a lot more detail has to be put into this simple statement. Even perhaps to the extent of a simple sketch or a list of criteria that the access scaffold must meet, for example

- minimum of two landings 1.2 × 2.4 m wide
- treads to be a minimum of 230 mm wide
- risers to be no more than 180 mm high
- staircase to be a minimum of 1.2 m wide
- safety handrail on both sides to be 1.2 m high, enclosed with mesh
- 30 lux lighting at 2 m centres vertically on scaffold.

Also required is a signing-off procedure with the first *Contractor* to hand to the next to prove that the staircase is indeed fit for its purpose.

Another argument is one where the *Contractor* gives notification that he has completed the shaft and the permanent works and he requests a Completion certificate from the *Project Manager*.

The *Project Manager* could answer that the *Contractor* has not put in the access stairs yet, therefore he (the *Project Manager*) will not give the *Contractor* a Completion certificate. The *Contractor* responds that the access stairs are temporary works, but the *Project Manager* advises the *Contractor* that as far as his *works* are concerned, they are temporary/permanent works which he must complete prior to a Completion certificate being issued.

If the *Project Manager* has not stipulated this requirement in the Works Information then an ambiguity/inconsistency exists which, when it is possible to apply the 'contra proferentem' rule that the ambiguity/inconsistency is construed against the drafter, then the *Project Manager* could lose his argument. (*Contra proferentem* means that the meaning of words works against the Party who drafted the words **or** who uses it as a basis for claim against another. For example, a plaintiff who sues for breach of a written contract can expect any ambiguity in the terms of the contract will be resolved against him.) If on the other hand it is clearly spelled out in the Works Information, then the *Project Manager* will be correct to insist on it being completed prior to his issuing a Completion certificate.

A key consideration in all interface management is safety. The project team should think about the likely problem areas with regard to health, safety, security, facilities, etc.

The access to the curtilage of the site is through a redundant petrol station and specifically the route into the site is over the redundant fuel tanks, which are covered in approximately 1 m of concrete.

On commencing the contract the *Contractor* notifies the *Project Manager* that he is proposing to bring onto site some heavy piling Equipment and other heavy Equipment, and wishes to have assurances that it is safe to bring the Equipment onto the Site.

Unfortunately no site investigation work has been undertaken on these 20-year-old tanks and the *Project Manager* faces the prospect of delays to the project, especially as the design team feel unable to provide the assurances sought by the *Contractor*. The *Project Manager* is simply unable to give assurances and therefore instructs the *Contractor* to investigate and then fill the previously de-gassed and cleaned tanks with Grade 40 Concrete. (This would be a compensation event as a change to the Works Information.)

The *Contractor* calls an early-warning meeting to notify the *Project Manager* that the reinforcement drawings for the new lift shaft call for 18 m-long 40 mm-diameter rebar to the waling beams. The new lift shaft is being constructed top down and while these bars can be placed in the ground-level waling beams the lower levels would require the bars to be lowered down vertically in a shaft which has men working below. This would be very dangerous, and very difficult to lower these heavy bars each weighing 200 kg safely. Therefore, the excavation work would have to stop while the rebar was lowered into the shaft. This stoppage in work had not been envisaged at time of tender and would delay the project.

The *Contractor* also pointed out that, once lowered, the rebar would be almost impossible to place in the walings.

The *Project Manager* agrees with the *Contractor* on both of these issues and instructs the *Contractor* to cut the rebar to lengths which would enable easy placing in the walings and the safe lowering of the rebar, while working continued at the bottom of the shaft. (This would be a compensation event as a change to the Works Information.)

## 4.8 General rules in drafting the Works Information

The contract places great emphasis on clarity in terms of contract documentation. An essential element of documentation is to present the documents in a clear and structured way. So at the outset careful consideration should be given to the structure and format of the contract documents.

### 4.8.1 Incorporating other documents into the Works Information

Where the Works Information is being drafted onto a blank sheet of paper, it is easier to incorporate the requirements of the ECC into the Works Information while leaving the sense of the document intact. Where a specification or *Employer*'s requirements is incorporated into the Works Information, problems could arise.

Where a standard specification or already-drafted specification is the basis of the Works Information, there are two options available when drafting the Works Information.

1 Incorporate the specification into the Works Information by reference; making sure that NEC terminology is adhered to, even if that is just by reading words in a different way. For example, a standard specification could be prefaced by an explanation of NEC terminology and could incorporate a table that includes items such as: 'Throughout this specification any reference to Engineer shall be deemed to mean *Project Manager*'.
This has inherent difficulties, however. The ECC *Project Manager* does not have exactly the same duties as the traditional Engineer. Some of the duties of the Engineer are those of the ECC *Project Manager*, and some are those of the ECC *Supervisor*. By automatically reading one term for another, ambiguities may be introduced into the document unnecessarily.
2 The preferred method of incorporating a specification into the Works Information is to rewrite the specification into the format of the Works Information, ensuring that all the requirements of the Works Information are catered for. (The rewrite could take the form of reading the specification from start to finish and manually making the changes required.) This is clearly more time-consuming, but achieves a more cohesive and comprehensive Works Information that should reduce the possibilities of disputes as a result of ambiguities and inconsistencies between documents. This proposed change also emphasises the importance of including sufficient time into the programme for the procurement of the project.

When compiling the Works Information the drafter needs to consider if the information he is providing is

- essential
- relevant

- complete
- clear
- unambiguous.

**4.8.2 General drafting rules**

In general, a few rules apply that assist in maintaining the cohesiveness of the Works Information.

1   Avoid words promoting conflict/adversarialism, such as
    - reasonable
    - to someone's satisfaction
    - suitable
    - appropriate
    - relevant.
2   Avoid legal and commercial terms. The appropriate place for legal and commercial terms is in the Contract Data.
3   Try to write a performance specification, particularly where the scope is not totally defined.
4   Documents are submitted for acceptance. The word 'approval' is not used in the NEC.
5   Do not rely on past relationships. The NEC is different from other forms of contract; so do not rely on the fact that the *Contractor* knows the way your company works.
6   Do not include any statements which refer to the tender. There should be no reference to the tender in any of the contract documents.

**4.8.3 The classic statements**

There are some classic 'woolly' statements with incorrect terminology, which appear in many specifications. A list of some of these classic statements is given below:

- 'The *works* will be complete when in the **opinion of the Engineer** all work is complete'.
- 'The *Contractor* shall have been deemed to have allowed in his tender for **all tests that the Engineer may require**'.
- 'Work will be completed to the **satisfaction of the** *Engineer*'.
- 'The *Contractor* shall allow for **all reasonable access** to the site'.
- 'The *Contractor* shall have been deemed to have visited the installation and to have **allowed for everything necessary to complete it**'.

If we look at the statements we can see that they lack clarity and are woolly in the sense that they do not spell out clearly and concisely what it is that the *Contractor* is to provide.

If he does not know what it is he is to provide how can he price for it? It also means that if a dispute occurs on an item there can be long arguments about what the clause actually means or was intended to mean rather than what it actually says.

Harry Haste, the *Project Manager*, is in a rush, his fees are tight and he sees no reason to reinvent the wheel. Consequently, he runs off copies of old specifications for the new contract. He is secure in the knowledge that his specification had always been very robust.

A typical example of his specification for the *works* follows.

**Examination of typical traditional specification:**

*W1   Water mains*
*W1.1   If, in the* opinion of the Engineer*, there is undue delay in the application of the first hydraulic test, or of any subsequent tests, or if any length of main should fail the test, the Engineer* may *direct the Contractor to suspend main laying operations until the length or lengths of main have been* satisfactorily *tested.*

[Note woolly words, 'in the opinion of' and 'satisfactorily', as well as the uncertainty regarding the Engineer's actions and the use of the term Engineer.]

W1.2 *After satisfactory completion of pressure testing, each valve on, and adjacent to, the pipeline shall be examined to witness that the opening/closing mechanisms function satisfactorily and are capable of functioning for their designated purpose. Isolating valves shall be examined by opening and closing the isolating gate three times. All tests and examinations shall be witnessed by the Engineer's staff and the Contractor shall provide a witnessed certificate to this effect to the Engineer.*

[Note woolly word 'satisfactorily' and the use of the wrong terminology. The witnessing of the tests should be by the *Supervisor*. If a preface had been included in this specification to the effect that read *Project Manager* for Engineer, then the *Project Manager* would have an obligation that is the *Supervisor*'s under the *conditions of contract* and a conflict would exist. Note also that the *Contractor* is required to provide notification of results under the *conditions of contract* and this Works Information paragraph requires an extra obligation of the *Contractor*.]

**W2      Swabbing of water mains**

W2.1 *Swabs will be provided by the Purchaser. Temporary pipework shall be provided by the Contractor.*

[Note wrong terminology in Purchaser, where it should refer to *Employer*. Note also the vague nature of the statement. How many swabs will be provided? Where will/can they be obtained from?]

**W3      Water supply for testing and swabbing**

W3.1 *For the purpose of the hydraulic testing of water-retaining structures and pipelines, the Purchaser will make available water from existing mains, at times and rates of flow to be decided by the Purchaser and agreed with the Engineer.*

[Note wrong terminology: should be *Project Manager* and *Employer*, not Purchaser and Engineer. Note also that if the *Contractor* were to do the test, he is not involved in the decision of when the water is to be provided, although this could affect his programme.]

W3.2 *Water will be provided free of charge to the Contractor for the first test. In the event of any part of the work having to be retested the Contractor shall be required to pay for the supply of water on a volume basis at the prevailing rate of the Purchaser.*

[Note wrong terminology: should be *Employer*, not Purchaser.]

**G1      Standards of materials**

G1.1 *15th Statement of the DoE Committee on Chemicals and Materials of Constructions for use in Public Water Supplies and Swimming Pools.*

G1.2 *The use of materials, including chemicals that do not meet the above criteria, shall be subject to the approval of the Engineer.*

[Note under an ECC contract 'acceptance' and not 'approval' is the terminology used. Also should be the *Project Manager*, not Engineer.]

**G2      Disinfection of pipework and structures**

*Pipes, pumps and structures shall be disinfected in accordance with the following procedure:*

(a) *The Contractor shall provide, at least three weeks before carrying out the disinfection process, a Method Statement to the Engineer for his Approval.*

[Note that in the ECC a method statement (or, in the case of ECC3, 'a statement of how the *Contractor* plans to do the work') is part of the *Contractor*'s programme. The *Contractor* submits his programme to the *Project Manager* (not the Engineer) for acceptance (not approval). There are several programme revisions provided for in the *conditions of contract*, including where the *Project Manager* can ask for one at any time.]

*Cont'd*

> *(b)*  *After <u>satisfactory</u> hydraulic testing, pipes, pumps, structures etc. shall be cleaned of all deleterious material.*
> [Note the use of vague words such as 'satisfactory'. The expected results of the test, as well as the details of how and when the test is to be carried out, should be provided.]
>
> *(c)*  *The volume to be disinfected shall be filled with chlorinated water at a dose level of 20–25 mg/l free chlorine and left to stand for 24 hours. Large structures may have all surfaces thoroughly scrubbed down with heavily chlorinated water and the volume then filled with a lower dose level, 0.5 mg/l chlorinated water, and left for 24 hours.*
>
> *(d)*  *The chlorinated water shall be drained away and disposed of in a <u>safe</u> and <u>satisfactory</u> manner. All <u>necessary</u> approvals for discharge shall be obtained by the Contractor. De-chlorination will be necessary to ensure that free chlorine discharged is below 0.1 mg/l, where discharge is to a watercourse, or drain leading to a watercourse.*
> [Note the vague words 'safe', 'satisfactory' and 'necessary'.
> What is safe and satisfactory to the *Contractor* could be different from the *Project Manager*. Unless a published standard is quoted, the requirements should be stated in the Works Information.]
>
> *(e)*  *The volume shall then be refilled with potable water and left a further 24 hours.*
>
> *( f )*  *<u>The Purchaser will take samples for bacteriological testing</u>. The results of his test will take a minimum of three working days to provide a conclusive test*
> [Note the incorrect terminology by the use of 'Purchaser' rather than *Employer*. The Works Information is required to state who provides samples, materials and facilities for testing. The criteria of the test should also be stated.]
>
> *(g)*  *<u>If test results are unsatisfactory to the Purchaser, the above procedure shall be repeated until satisfactory results are achieved.</u>*
> [Note: use of incorrect terminology 'Purchaser' and 'satisfactory'. The uncertainty of this statement could leave the *Contractor* having difficulties pricing accurately, or potentially adding conservative amounts of risk into his price.]

This emphasises the importance that the contract puts upon the Works Information.

**4.8.4 Conclusion**

There can be a world of difference between what the **intent** of a clause was and what it actually states. Great care is required in the writing and drafting of specifications.

Many people do not realise the importance of the Works Information and consequently more often than not the specifications remain very much like those found in JCT or ICE Contracts. Far too often old or similar documents are used with just a few words changed here and there. This is totally unsatisfactory. Little care or forethought is given to them and the seeds of dispute have inadvertently been set.

**4.9 Site Information**

**4.9.1 Introduction**

'Site Information is information which

- describes the Site and its surrounding and
- is in the documents which the Contract Data states it is in.'

Site information will cover items such as

- existing buildings on or adjacent to the Site
- existing mains services
- access for inspection of Site and buildings.

**4.9.2 Using the Site Information**

The Site Information has particular impact when the compensation event for physical conditions (clause 60.1(12)) is considered. In assessing the risk under the contract, the *Contractor* will take into account the Site Information (clause 60.2).

Those *Employers* who equate clause 60.1(12) with clause 12 of the ICE conditions may choose to 'pass the risk' in clause 60.1(12) to the *Contractor*, as was so frequently done with ICE clause 12. It should be noted, however, that risks are not 'passed' but are reallocated. The *Employer* is likely still to have to pay for that risk in one form or other, even if it is not as visible, for example, when the *Contractor* includes risk in his price, but that amount of risk cannot be separated from the rest of the price.

If an *Employer* does choose to delete clause 60.1(12) from the *conditions of contract* using Option Z, then the Site Information, for example, geological surveys, should not form part of the contract documents but should be made available as reference information about the Site and its surroundings. Only factual information should be made available, with the interpretation of the information being left to the *Contractor*. In addition, all references to Site Information should be removed from the documents, as should clauses 60.2 and 60.3 of the *conditions of contract*.

### 4.9.3 Impact of the Site Information for interfacing *Contractors*

Let us consider an example scenario.

The *Employer* has carried out some enabling works; part of that is the site establishment of the work site known as Spring Field. The plan layout shown on the Site Information for the follow-on *Contractor* is as shown in Figure 4.2.

The drawing issued to the follow-on *Contractor* shows the working area to be as shown on drawing reference 100 Spring Field Dell Site Rev A.

The *Contractor* uses this Site Information in the preparation of his tender. When he arrives on site he finds that the area of the site is somewhat different to that anticipated. The recourse he has is through compensation event 60.1(12) where the physical conditions affect the *Contractor*, taking into account the Site Information. The third bullet point of clause 60.1(12) requires the *Contractor* to have visited the Site and done what an experienced contractor would do. It is likely that the *Employer* would rely on this in any defence against the compensation event. The *Contractor* should always ensure that he has visited the Site and done everything possible to ensure that he is familiar with the Site and its conditions.

This example indicates the great care required when dealing with interface information between *Contractors*.

With the advent of the CDM Regulations many arguments can arise due to disagreement on how a Site has been left by the previous *Contractor* and how the following *Contractor* was advised he would find the Site.

**Figure 4.2** Site Information

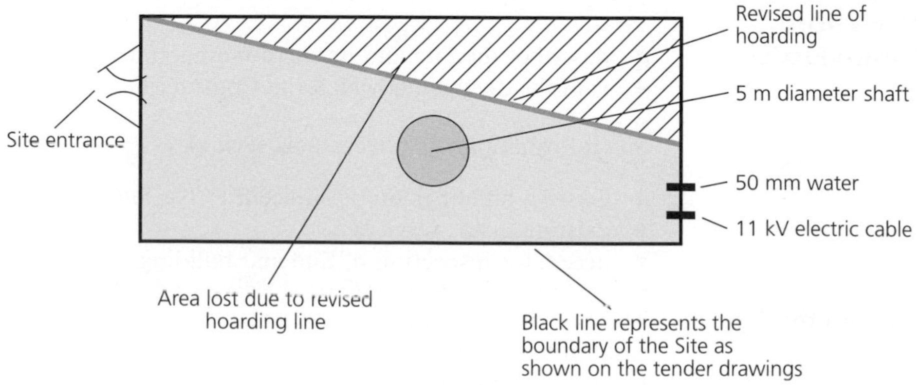

Site entrance

Revised line of hoarding

5 m diameter shaft

50 mm water

11 kV electric cable

Area lost due to revised hoarding line

Black line represents the boundary of the Site as shown on the tender drawings

### 4.9.4 Drawings issued for information purposes only

If an *Employer* issues drawings 'for information purposes only', what is the real status of that drawing? What if that information conflicts with other information provided?

In accordance with clause 60.2, the *Employer* takes the responsibility for information provided in the Site Information. The *Employer* cannot shrug off that responsibility by saying that the drawing was provided 'for information purposes only'. It remains his responsibility and he is liable under the contract if that information is incorrect or conflicting.

If the information is conflicting, then the *contra proferentem* rule (included in clause 60.3) applies and the information is interpreted in favour of the *Contractor* (since it was the *Employer* who drafted it).

This emphasises that the information provided in the Site Information is extremely important and cannot just be thrown together in a haphazard fashion. The *Employer* should ensure that it makes sense and that it represents accurately the Site because the *Contractor* relies on this information in putting together his tender. (See Figure 4.3.)

**Figure 4.3** Information purposes only

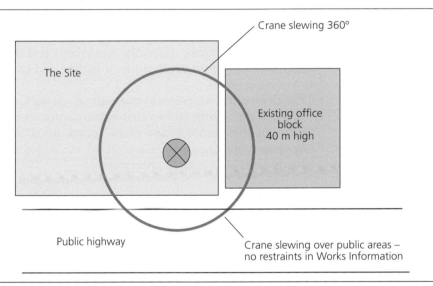

### 4.9.5 The Site, *boundaries of the site* and Working Areas

It is important to understand the terminology used in the contract. This section looks at the difference between the terms

- the Site
- *boundaries of the site*
- Working Areas.

### 4.9.5.1 The Site

The Site is defined as the area within the *boundaries of the site* stated in the Contract Data. This area comprises locations provided by the *Employer* for the *works*. Definition of the Site is important as it relates to

- *access dates*
- compensations events; for example, the *Employer* fails to give access to the Site by the date shown on the Accepted Programme
- title to objects/items within the Site
- definition of *Employer*'s risks; that is, 'use or occupation of the Site'
- termination – *Employer* may instruct the *Contractor* to leave the Site.

### 4.9.5.2 The boundaries of the site

The Site is defined as the area within the *boundaries of the site* (stated in the Contract Data). This area comprises locations provided by the *Employer* for the *works*. It is usually

**Figure 4.4** The Site, *boundaries of the site* and Working Areas

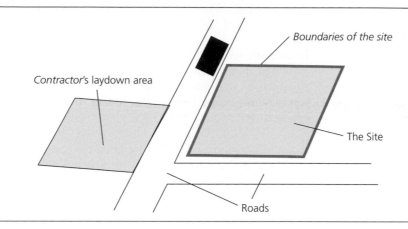

identified in Contract Data part one by the *Employer* by naming a drawing on which the Site can be identified, and perhaps outlining in highlighter the *boundaries of the site* (see Figure 4.4).

**4.9.5.3 The Working Areas**

The concept of Working Areas is introduced in recognition of the fact that the *Contractor* often makes use of other areas to provide the *works*. Such areas may include, batching plant, laydown areas, temporary workshops and the like. For example, in Figure 4.4 the *Employer* has stated the location of the site and its *boundaries*.

The *Contractor*, when completing part two of the Contract Data, might decide that the area of the Site is insufficient to enable him to prefabricate piling reinforcement cages for the new building and he also plans to set up a temporary bar reinforcement bending workshop. By chance there is a vacant site across the road and adjacent to the site which will be suitable for his requirements.

The *Contractor* starts negotiations with the landowner – which he knows he will not have concluded prior to him submitting his tender. He therefore inserts in part two of the Contract Data, not the specific site name, but a more generic one which simply states 'laydown and prefabrication facilities'.

Alternatively, it could also transpire that during the hurly-burly of preparing his tender for submission, he forgets to put into the tender the other *working areas* that he requires in order to Provide the Works. He only discovers this omission after the contract has been let. In this case, the *Contractor* should write to the *Project Manager* under clause 15.1 submitting a proposal for adding to the Working Areas and stating his reasons why he requires the additional area. As long as the proposal is necessary for providing the *works* and it will be used for work, which is in the contract, the *Project Manager* should have no objection.

Working Areas has significance in that under clause 70.2 the title that the *Contractor* has in Plant and Materials brought within the Working Areas passes to the *Employer*. Title to Plant and Materials passes back to the *Contractor* if it is removed from the Working Area with the *Project Manager*'s permission. Working Areas is discussed in greater detail in Chapter 2 of Book 4, which deals with the Schedule of Cost Components, since the Working Areas' identification impacts on Defined Cost.

**Procuring an Engineering and Construction Contract**
ISBN 978-0-7277-5720-3

ICE Publishing: All rights reserved
doi: 10.1680/pecc.57203.139

# Appendix 4
# Works Information clauses

## A4.1 Works Information clauses

The following are clauses that appear in the *conditions of contract* and that require further information to be inserted into the Works Information.

| | |
|---|---|
| 11.2(2) | Completion is when the *Contractor* has<br>■ done all the work which the Works Information states he is to do by the Completion Date and<br>■ corrected notified Defects which would have prevented the *Employer* and Others from doing their work. |
| 11.2(5) | A Defect is<br>■ a part of the *works* which is not in accordance with the Works Information or<br>■ a part of the *works* designed by the *Contractor* which is not in accordance with the applicable law or the *Contractor*'s design which the *Project Manager* has accepted. |
| 11.2(7) | Equipment is items provided by the *Contractor* and used by him to Provide the Works and which the Works Information does not require him to include in the *works*. |
| 11.2(13) | To Provide the Works means to do the work necessary to complete the *works* in accordance with this contract and all incidental work, services and actions which this contract requires. |
| 11.2(19) | Works Information is information which either<br>■ specifies and describes the *works* or<br>■ states any constraints on how the *Contractor* Provides the Works<br>and is either<br>■ in the documents which the Contract Data states it is in or<br>■ in an instruction given in accordance with this contract. |
| 18.1 | The Contractor notifies the *Project Manager* as soon as he considers that the Works Information requires him to do anything which is illegal or impossible. If the *Project Manager* agrees, he gives an instruction to change the Works Information appropriately. |
| 20.1 | The *Contractor* Provides the Works in accordance with the Works Information. |
| 21.1 | The *Contractor* designs the parts of the *works* which the Works Information states he is to design. |
| 21.2 | The *Contractor* submits the particulars of his design as the Works Information requires to the *Project Manager* for acceptance. A reason for not accepting the *Contractor*'s design is that<br>■ it does not comply with the Works Information or<br>■ it does not comply with the applicable law.<br><br>The *Contractor* does not proceed with the relevant work until the *Project Manager* has accepted his design. |
| 22.1 | The *Employer* may use and copy the *Contractor*'s design for any purpose connected with construction, use, alteration or demolition of the *works* unless otherwise stated in the Works Information and for other purposes as stated in the Works Information. |
| 25.1 | The *Contractor* cooperates with Others in obtaining and providing information which they need in connection with the *works*. He cooperates with Others and shares the Working Areas with them as stated in the Works Information. |
| 25.2 | The *Employer* and the *Contractor* provide services and other things as stated in the Works Information. Any cost incurred by the *Employer* as a result of the *Contractor* not providing the services and other things which he is to provide is assessed by the *Project Manager* and paid by the *Contractor*. |
| 27.4 | The *Contractor* acts in accordance with the health and safety requirements stated in the Works Information. |

| | |
|---|---|
| 31.2 | The *Contractor* shows on each programme which he submits for acceptance<br>• the *starting date*, *access dates*, Key Dates and Completion Date<br>• planned Completion<br>• the order and timing of the operations which the *Contractor* plans to do in order to Provide the Works<br>• the order and timing of work of the *Employer* and Others as last agreed with them by the *Contractor* or, if it not so agreed, as stated in the Works Information,<br>• the dates when the *Contractor* plans to meet each Condition stated for the Key Dates and to complete other work needed to allow the *Employer* and Others to do their work,<br>• provisions for<br>  • float,<br>  • time risk allowances,<br>  • health and safety requirements and<br>  • the procedures set out in this contract,<br>• the dates when, in order to Provide the Works in accordance with his programme, the *Contractor* will need<br>  • access to a part of the Site if later than its *access date*,<br>  • acceptances<br>  • Plant and Materials and other things to be provided by the *Employer* and<br>  • information from Others,<br>• for each operation, a statement of how the *Contractor* plans to do the work identifying the principal Equipment and other resources which he plans to use and<br>• other information which the Works Information requires the *Contractor* to show on a programme submitted for acceptance. |
| 31.3 | Within two weeks of the *Contractor* submitting a programme to him for acceptance, the *Project Manager* either accepts the programme or notifies the *Contractor* of his reasons for not accepting it. A reason for not accepting a programme is that<br>• the *Contractor*'s plans which it shows are not practicable,<br>• it does not show the information which this contract requires,<br>• it does not represent the *Contractor*'s plans realistically or<br>• it does not comply with the Works Information. |
| 35.2 | The *Employer* may use any part of the *works* before Completion has been certified. If he does so, he takes over part of the *works* when he begins to use it, except if the use is<br>• for a reason stated in the Works Information or<br>• to suit the *Contractor*'s method of working. |
| 40.1 | The subclauses in this clause only apply to tests and inspections required by the Works Information or the applicable law. |
| 40.2 | The *Contractor* and the *Employer* provide materials, facilities and samples for tests and inspections as stated in the Works Information. |
| 41.1 | The *Contractor* does not bring to the Working Areas those Plant and Materials which the Works Information states are to be tested or inspected before delivery, until the *Supervisor* has notified the *Contractor* that they have passed the test or inspection. |
| 42.1 | Until the *defects date*, the *Supervisor* may instruct the *Contractor* to search for a Defect. He gives his reason for the search with the instruction. Searching may include<br>• uncovering, dismantling, re-covering and re-erecting work<br>• providing facilities, materials and samples for tests and inspections done by the *Supervisor* and<br>• doing tests and inspections which the Works information does not require. |

| 60.1(5) | The *Employer* or Others |
|---|---|
| | ■ do not work within the times shown on the Accepted Programme |
| | ■ do not work within the conditions stated in the Works Information or |
| | ■ carry out work on the Site that is not stated in the Works Information. |
| 60.1(16) | The *Employer* does not provide materials, facilities and samples for tests and inspections as stated in the Works Information. |
| 71.1 | The *Supervisor* marks Equipment, Plant and Materials which are outside the Working Areas if |
| | ■ this contract identifies them for payment and |
| | ■ the *Contractor* has prepared them for marking as the Works Information requires. |
| 73.2 | The *Contractor* has title to materials from excavation and demolition only as stated in the Works Information. |
| A, C54.1 | Information in the Activity Schedule is not Works Information or Site Information. |
| B, D55.1 | Information in the Bill of Quantities is not Works Information or Site Information. |
| C, D, E11.2(25) and F11.2(26) | Disallowed Cost is cost which the *Project Manager* decides |
| | ■ is not justified by the *Contractor*'s accounts and records, |
| | ■ should not have been paid to a Subcontractor or supplier in accordance with his contract, |
| | ■ was incurred only because the *Contractor* did not |
| |     ■ follow an acceptance or procurement procedure stated in the Works Information or |
| |     ■ give an early warning which this contract required him to give |
| | and the cost of |
| | ■ correcting Defects after Completion |
| | ■ correcting Defects caused by the *Contractor* not complying with a constraint on how he is to Provide the Works stated in the Works Information |
| | ■ Plant and Materials not used to Provide the Works (after allowing for reasonable wastage) unless resulting from a change to the Works Information |
| | ■ resources not used to Provide the Works (after allowing for reasonable availability and utilisation) or not taken away from the Working Areas when the *Project Manager* requested and |
| | ■ preparation for and conduct of an adjudication or proceedings of the *tribunal*. |
| | (Note that F11.2(26) is slightly different but still contains a reference to the Works Information.) |
| C, D, E, F52.2 | The *Contractor* keeps these records |
| | ■ accounts of his payments of Defined Cost, |
| | ■ proof that the payments have been made, |
| | ■ communications about and assessments of compensation events for Subcontractors and |
| | ■ other records as stated in the Works Information. |
| X4.1 | If a parent company owns the *Contractor*, the *Contractor* gives to the *Employer* a guarantee by the parent company of the *Contractor*'s performance in the form set out in the Works Information. If the guarantee was not given by the Contract Date, it is given to the *Employer* within four weeks of the Contract Date. |
| X13.1 | The *Contractor* gives the *Employer* a performance bond, provided by a bank or insurer which the *Project Manager* has accepted, for the amount stated in the Contract Data and in the form set out in the Works Information. A reason for not accepting the bank or insurer is that its commercial position is not strong enough to carry the bond. If the bond was not given by the Contract Date, it is given to the *Employer* within four weeks of the Contract Date. |

| | |
|---|---|
| X14.2 | The advance payment is made either within four weeks of the Contract Date or, if an advanced payment bond is required, within four weeks of the later of<br>■ the Contract Date and<br>■ the date when the *Employer* receives the advanced payment bond.<br>The advance payment bond is issued by a bank or insurer which the *Project Manager* has accepted. A reason for not accepting the proposed bank or insurer is that its commercial position is not strong enough to carry the bond. The bond is for the amount of the advanced payment which the *Contractor* has not repaid and is in the form set out in the Works Information. Delay in making the advanced payment is a compensation event. |
| X15.1 | The *Contractor* is not liable for Defects in the *works* due to his design so far as he proves that he used reasonable skill and care to ensure that his design complied with the Works Information |

**Procuring an Engineering and Construction Contract**
ISBN 978-0-7277-5720-3

ICE Publishing: All rights reserved
doi: 10.1680/pecc.57203.145

# Index

accepted programme, 62, 93, 137, 141
access dates, 137, 140, 141
activity schedule, 19, 21, 25, 26, 29, 38, 40, 42, 43–45, 46, 47, 49, 51–53, 63, 69, 87, 109
   activities designed to suit assessment dates, figure, 45
   activities spanning assessment date, figure, 44
   activities too small to be included in programme, figure, 44
   example, 35
adjudicator, 15, 60, 67, 97
adjudicator contract, 60
adjudicator nominating body, 60, 104
administration, 15
advanced payment bond, 63–64, 121, 142
approval, 124, 127
arbitration, 15, 60, 96, 104
assessment date, 43–44, 45, 94, 106
audit, 11, 57, 59, 69–83
   charges, 70, 78–80, 80–81
   design, 70, 81–83
   equipment, 70, 75–77
   goals, 70
   information, provided by project team, 75, 77, 78, 80, 83
   manufacture and fabrication, 70, 80–81
   objectives, 69–70
      charges, 79
         design, 81–82
         examination of consistency, 81
         examination of payslips, productivity basis, 81
         examination of payslips, time basis, 81
         spot checks on labour costs, 81
      equipment, 75–76
         pre-order and post-order, 75
      manufacture and fabrication, 80
         examination of build-up to payments, 80
         visit site of manufacture and fabrication, 80
      people, 74
      plant and materials, 77–78
         pre-order and post-order, 77
      subcontractors, 70, 72, 73
         pre-award and post award, 72
   people, 74–75
   plan, 69
      amendments, 70
      basis of plan and procedures, 69
      distribution, 70
   plant and materials, 70, 76, 77–78
   procedures
      charges, 79
      design, 81–82
      equipment, 75–76
      manufacture and fabrication, 80
      plant and materials, 77–78
      subcontractors, 72–73
   records
      charges, 79–80, 81
      design, 82
      equipment, 76
      manufacture and fabrication, 80–81
      people, 74–75
      plant and materials, 78
      subcontractors, 73–74
   reports, 70
      charges, 79, 80
      design, 82
      equipment, 76
      manufacture and fabrication, 80
      people, 74, 75
      plant and materials, 78
      subcontractors, 74
   reporting, 70
      comments sheet, 71
   subcontractors, 70, 72–74

back to back contracts, 40
base date index, 61
bill of quantities, 19, 25, 26, 38, 40, 41, 42, 45–46, 49, 53, 63, 69, 87, 103, 109, 110
   example, 35
bonds, 60, 63–64, 60, 63, 86, 88, 121, 142
bonus for early completion, 62, 86
boundaries of the site, 137–138

cash flow, 44, 45
charges, 70, 78–80
Civil Engineering Contractors Association (CECA) Dayworks Schedule, 110, 113
clause 20.3, 55
clause 20.4, 55–56
collateral warranties, 25, 34
commercial aspects, 24–25, 26, 27
common law, 67
compensation event, 15, 20, 26, 28, 42, 43, 59, 61, 64, 90, 100–101, 108, 113, 116, 122, 124, 130, 136, 142
   activity schedule, 44–45, 52
   audit, 72
   bill of quantities, 46
   *see also* damages
completion, 21, 22, 49, 52, 55, 62, 64, 65, 89, 105, 107, 120–121, 140, 142
completion, early, 62, 86, 90, 94
completion, sectional, 61–62, 86